AF556186

Thermodynamics and Statistical Mechanics

Thermodynamics and Statistical Mechanics

Er. Devesh Kumar Dixit

RANDOM PUBLICATIONS
NEW DELHI (INDIA)

Thermodynamics and Statistical Mechanics

ISBN 978-93-5111-255-6

Published in 2014 in India by

RANDOM PUBLICATIONS

4376-A/4B, Gali Murari Lal, Ansari Road
New Delhi-110 002
Phone : +91-11-43580356, +91-11-23289044
e-mail: randomexports@gmail.com, sales@randompublications.com, info@randompublications.com

Reprinted 2021

Type Setting by : Keystoneprintads, Delhi-110051
Digitally Printed at: Replika Press Pvt. Ltd.

Preface

Thermodynamics is a branch of physics which deals with the energy and work of a system. It was born in the 19th century as scientists were first discovering how to build and operate steam engines. Thermodynamics deals only with the large scale response of a system which we can observe and measure in experiments.

Small scale gas interactions are described by the kinetic theory of gases. The methods complement each other; some principles are more easily understood in terms of thermodynamics and some principles are more easily explained by kinetic theory.

Thermodynamics is basic science that deals with energy and has long been an essential part of engineering curricula all over the world. It was developed during the eighteenth and nineteenth century during a time when temperature and heat were not well understood yet.

Its development was driven by the need for an improved theoretical understanding of steam engines invented at about the same time. It evolved as a rather formal and elegant theory that proved to be of great importance to engineers.

Statistical mechanics is the application of statistics, which includes mathematical tools for dealing with large populations, to the field of mechanics, which is concerned with the motion of particles or objects when subjected to a force.

It provides a framework for relating the microscopic properties of individual atoms and molecules to the macroscopic or bulk properties of materials that can be observed in every day life, therefore explaining thermodynamics as a natural result of statistics and mechanics.

This book will be of use to wide audience of students and professionals in the fields of chemistry, chemical engineering and physics.

I thank all members of my team who have helped in the preparation of the book. My special thanks go to "Random Publications" who have published the book.

–Er. Devesh Kumar Dixit

Contents

Preface .... *v-vi*

1. **Introduction to Thermodynamics** **1**

Application of First Law of Thermodynamics 3
The Second Law of Thermodynamics 6
The Third Law of Thermodynamics 7
Ideal Gases 10
Ideal Gas Model of Thermodynamics 16

2. **Boltzmann Distributions and Thermodynamics** **33**

Boltzmann Distributions 33
Simplest Microscopic System 36
Thermodynamic Relations 43
The Maxwell Distribution 48

3. **Thermodynamic Systems** **54**

Closed and Open System of Thermodynamic System 55
Effects of Flow and Non-flow Processes 58
Thermodynamic Parameters 65
Classical Thermodynamics 67
Types of Thermodynamic Processes 76
Definition of Thermodynamic State 79
Thermodynamics System 82

4. **The Equipartition Theorem** **86**

Internal Energy of a Monatomic Ideal Gas 86
The Classical Approximation of Equipartition Theorem 88
Entropies of Substances 90

Investigate the Specific Heats of Gases 92

5. Entropy 105

Introduction 105
Change in Entropy 108
Entropy System 115
Entropy Message 125
Entropy in Quantum Mechanics 135
Microscopic Definition of Entropy (Statistical Mechanics) 143

6. Laws of Thermodynamics 147

Introduction 147
Zeroth Law of Thermodynamics 147
Laws of Conservation of Energy 150
First Law of Thermodynamics 151
Second Law of Thermodynamics 160
The Third Law of Thermodynamics 168

7. Heat Engines 172

Introduction 172
Heat 178
Heat Capacities 182
Heat Engines and Refrigerators 190
Second Law of Thermodynamics 192
Practical Engines 198
Specific Heat Capacities of Ideal Gas 207
P – V –T Surface of an Ideal Gas 208

8. Engineering Thermodynamics 210

Definition of Engineering Thermodynamics 210
Applications of Engineering Thermodynamics 211
An Idealization of Thermodynamic Cycles 215

9. Statistical Thermodynamics 223

Mechanical Interaction and Macrosystems 223
Temperature 227
Statistical Mechanics 230
Investigation of Statistical Thermodynamics 234
Thermodynamic Properties of Platinum Diatomics 237

10. Elementary Concepts in Statistics 240

Fundamentals of Statistics 240
Elementary Statistical Concepts 246

Mathematical Statistics 250
Study of Statistical Ensembles of Quantum Mechanical Systems 251
Uses of Statistics 255
Power Law Distribution in Statistical Mechanics 257

11. The Ideal Gas Law and Kinetic Theory 262

The Mole, Avogadro's Number, and Molecular Mass 262
Ideal Bose gas 263
The Ideal Gas Law 263
Ideal Fermi gas 264
Comparison of Solids, Liquids, and Gases 266
The Gas Specific Heats CV and CP 269
Specific Heat of an Ideal Gas 277
Kinetic Energy and Mass for Very Fast Particles 279

***Index* 283**

1

Introduction to Thermodynamics

Thermodynamics is a branch of physics which deals with the energy and work of a system. It was born in the 19th century as scientists were first discovering how to build and operate steam engines. Thermodynamics deals only with the large scale response of a system which we can observe and measure in experiments. Small scale gas interactions are described by the kinetic theory of gases. The methods complement each other; some principles are more easily understood in terms of thermodynamics and some principles are more easily explained by kinetic theory.

There are three principal laws of thermodynamics which are described on separate slides. Each law leads to the definition of thermodynamic properties which help us to understand and predict the operation of a physical system. We will present some simple examples of these laws and properties for a variety of physical systems, although we are most interested in thermodynamics in the study of propulsion systems and high speed flows. Fortunately, many of the classical examples of thermodynamics involve gas dynamics. Unfortunately, the numbering system for the three laws of thermodynamics is a bit confusing. We begin with the zeroth law.

The zeroth law of thermodynamics involves some simple definitions of thermodynamic equilibrium. Thermodynamic equilibrium leads to the large scale definition of temperature, as opposed to the small scale definition related to the kinetic energy of the molecules. The first law of thermodynamics relates the various forms of kinetic and potential energy in a system to the work which a system can perform and to the transfer of heat. This law is sometimes taken as the definition of internal energy, and introduces an additional state variable, enthalpy. The first law of thermodynamics allows for many possible states of a system to exist. But experience indicates that only certain states occur. This leads to the second law of thermodynamics and the definition of another state variable called entropy. The second law stipulates that the total entropy of a system plus its environment can not decrease; it can remain constant for a reversible process but must always increase for an irreversible process.

THERMODYNAMIC POTENTIALS

As can be derived from the energy balance equation on a thermodynamic system there exist energetic quantities called thermodynamic potentials, being the quantitative measure of the stored energy in the system. The five most well known potentials are:

- Internal energy U
- Helmholtz free energy $A = U - TS$
- Enthalpy $H = U + PV$
- Gibbs free energy $G = U + PV - TS$
- Grand potential $\Phi_G = U - TS - \mu N$

Other thermodynamic potentials can be obtained through Legendre transformation. Potentials are used to measure energy changes in systems as they evolve from an initial state to a final state. The potential used depends on the constraints of the system, such as constant temperature or pressure. Internal energy is the internal energy of the system, enthalpy is the internal energy of the system plus the energy related to pressure-volume work, and Helmholtz and Gibbs energy are the energies available in a system to do useful work when the temperature and volume or the pressure and temperature are fixed, respectively.

CLASSICAL THERMODYNAMICS

Classical thermodynamics is the original early 1800s variation of thermodynamics concerned with thermodynamic states, and properties as energy, work, and heat, and with the laws of thermodynamics, all lacking an atomic interpretation. In precursory form, classical thermodynamics derives from chemist Robert Boyle's 1662 postulate that the pressure P of a given quantity of gas varies inversely as its volume V at constant temperature; i.e. in equation form: $PV = k$, a constant. From here, a semblance of a thermo-science began to develop with the construction of the first successful atmospheric steam engines in England by Thomas Savery in 1697 and Thomas Newcomen in 1712. The first and second laws of thermodynamics emerged simultaneously in the 1850s, primarily out of the works of William Rankine, Rudolf Clausius, and William Thomson.

STATISTICAL THERMODYNAMICS

With the development of atomic and molecular theories in the late 1800s and early 1900s, thermodynamics was given a molecular interpretation. This field is called statistical thermodynamics, which can be thought of as a bridge between macroscopic and microscopic properties of systems. Essentially, statistical thermodynamics is an approach to thermodynamics situated upon statistical mechanics, which focuses on the derivation of macroscopic results from first principles. It can be opposed to its historical predecessor phenomenological thermodynamics, which gives scientific descriptions of phenomena with avoidance of microscopic details. The statistical approach is to derive all macroscopic properties (temperature, volume, pressure, energy, entropy, etc.) from the properties of moving constituent particles and the interactions between them (including quantum phenomena). It was found to be very successful and thus is commonly used.

CHEMICAL THERMODYNAMICS

Chemical thermodynamics is the study of the interrelation of heat with chemical reactions or with a physical change of state within the confines of the laws of thermodynamics. On the Equilibrium of Heterogeneous Substances, in which he showed how thermodynamic processes could be graphically analyzed, by studying the energy, entropy, volume, temperature and pressure of the thermodynamic system, in such a manner to determine if a process would occur spontaneously. During the early 20th century, chemists such as Gilbert N. Lewis, Merle Randall, and E.A. Guggenheim began to apply the mathematical methods of Gibbs to the analysis of chemical processes.

APPLICATION OF FIRST LAW OF THERMODYNAMICS

Entropy is a function of a quantity of heat which shows the possibility of conversion of that heat into work. For a thermodynamic system with a fixed number of particles, the first law of thermodynamics may be stated as:

$$\delta Q = dU + \delta W \text{, or equivalently, } dU = \delta Q - \delta W,$$

where δQ is the amount of energy added to the system by a heating process, δW is the amount of energy lost by the system due to work done by the system on its surroundings and dU is the increase in the internal energy of the system.

The δ's before the heat and work terms are used to indicate that they describe an increment of energy which is to be interpreted somewhat differently than the dU increment of internal energy. Work and heat are processes which add or subtract energy, while the internal energy U is a particular form of energy associated with the system. Thus the term "heat energy" for δQ means "that amount of energy added as the result of heating" rather than referring to a particular form of energy. Likewise, the term "work energy" for δW means "that amount of energy lost as the result of work". The most significant result of this distinction is the fact that one can clearly state the amount of internal energy possessed by a thermodynamic system, but one cannot tell how much energy has flowed into or out of the system as a result of its being heated or cooled, nor as the result of work being performed on or by the system. In simple terms, this means that energy cannot be created or destroyed, only converted from one form to another. For a simple compressible system, the work performed by the system may be written

$$\delta Q = P\,dV,$$

where P is the pressure and dV is a small change in the volume of the system, each of which are system variables. The heat energy may be written

$$\delta Q = T\,dS,$$

where T is the temperature and dS is a small change in the entropy of the system. Temperature and entropy are also system variables.

MECHANICS

In mechanics, conservation of energy is usually stated as

$$E = T + V,$$

where T is kinetic and V potential energy.

Actually this is the particular case of the more general conservation law

$$\sum_{i=1}^{N} p_i \dot{q}_i - L = const \text{ and } p_i = \frac{\partial L}{\partial \dot{q}_i}$$

where L is the Lagrangian function. For this particular form to be valid, the following must be true:

- The system is scleronomous (neither kinetic nor potential energy are explicit functions of time)
- The kinetic energy is a quadratic form with regard to velocities.
- The potential energy doesn't depend on velocities.

Noether's Theorem

In many physical theories. It is understood as a consequence of Noether's theorem, which states every symmetry of a physical theory has an associated conserved quantity; if the theory's symmetry is time invariance then the conserved quantity is called "energy". In other words, if the theory is invariant under the continuous symmetry of time translation then its energy (which is canonical conjugate quantity to time) is conserved.

Conversely, theories which are not invariant under shifts in time (for example, systems with time dependent potential energy) do not exhibit conservation of energy — unless we consider them to exchange energy with another, external system so that the theory of the enlarged system becomes time invariant again. Since any time-varying theory can be embedded within a time-invariant meta-theory energy conservation can always be recovered by a suitable re-definition of what energy is. Thus conservation of energy is valid in all modern physical theories, such as special and general relativity and quantum theory (including QED).

Relativity

With the invention of special relativity by Albert Einstein, energy was proposed to be one component of an energy-momentum 4-vector. Each of the four components (one of energy and three of momentum) of this vector is separately conserved in any given inertial reference frame. Also conserved is the vector length (Minkowski norm), which is the rest mass. The relativistic energy of a single massive particle contains a term related to its rest mass in addition to its kinetic energy of motion. In the limit of zero kinetic energy (or equivalently in the rest frame of the massive particle, or the center-of-momentum frame for objects or systems), the total energy of particle or object (including internal kinetic energy in systems) is related to its rest mass via the famous equation E = mc2.

Thus, the rule of conservation of energy in special relativity was shown to be a special case of a more general rule, alternatively called the conservation of mass and energy, the conservation of mass-energy, the conservation of energy-momentum, the conservation of invariant mass or now usually just referred to as conservation of energy. In general relativity conservation of energy-momentum is expressed with the aids of a stress-energy-momentum pseudotensor.

Quantum Theory

In quantum mechanics, energy is defined as proportional to the time derivative of the wave function. Lack of commutation of the time derivative operator with the time operator itself mathematically results in an uncertainty principle for time and energy: the longer the period of time, the more precisely energy can be defined (energy and time become a conjugate Fourier pair). However, there is a deep contradiction between quantum theory's historical estimate of the vacuum energy density in the universe and the vacuum energy predicted by the cosmological constant.

The estimated energy density difference is of the order of 10^120 times. The consensus is developing that the quantum mechanical derived zero-point field energy density does not conserve the total energy of the universe, and does not comply with our understanding of the expansion of the universe. Intense effort is going on behind the scenes in physics to resolve this dilemma and to bring it into compliance with an expanding universe.

Mathematical Viewpoint

From a mathematical point of view, the energy conservation law is a consequence of the shift symmetry of time; energy conservation is implied by the empirical fact that the laws of physics do not change with time itself. Philosophically this can be stated as "nothing depends on time per se".

THE LAW OF CONSERVATION OF MATTER/ENERGY

The first law has been defined as follows: When a closed system is altered adiabatically, the total work associated with the change of state is the same for all possible processes between the two given equilibrium states. A more succinct and comprehensible definition might be something like this: Matter/energy may be altered but not created (from nothingness) nor destroyed (reduced to nothingness). The First Law teaches that matter/energy cannot spring forth from nothing without cause, nor can it simply vanish.

The First Law, although not formally defined until the 19th century, helps make science possible. Science depends on the ability to identify cause-effect relationships. If matter/energy could spontaneously appear (and have effects on other matter/energy around it), scientists would never know whether a given observation was due to a rational cause, or to a spontaneous generation of matter or energy that was uncaused. Scientific conclusions would be on shaky ground. The Law of Causality is thus closely linked with the First Law of Thermodynamics.

The First Law also demands, if we accept it, one of two possibilities about the nature of the universe. One is that it has always existed, changing form perhaps but never having come from nothingness, or returning to the same. The other possibility is that it did not come from nothingness, but from a transcendant (that is, outside the universe) creator who is not subject to the laws within the universe.

First, they have unconsciously granted to the Law of Causality the very property of self-existence (that is, an eternal, uncreated nature) that they are presuming God couldn't have. A being who created the universe and the laws within it, who pre-existed them, would not be slave to those laws. And since the Law of Causality is a statement about relationships between multiple entities, the law could not even exist until one entity began the act of creating another one (at which point it would implicitly come to exist).

Finally, most atheists who use this argument grant to the universe the exact property of self-existence that they deny God. They either deny the First Law of Thermodynamics and believe the universe came into existence from nothingness, or believe that it is itself self-existing. However, this latter position violates the unity principle – that a valid law of science that is found to apply anywhere, applies everywhere and to everything in the universe, including the universe as a whole.

The only position that appears to be consistent with the First Law of Thermodynamics, the unity principle and causality is that the universe was created by a self-existent external agent not subject to the laws operational in the universe it created.

THE SECOND LAW OF THERMODYNAMICS

In plain English the Second Law states that entropy always increases or remains constant in a closed system. (As a practical matter, for any non-trivial system entropy tends to increase due to irreversible processes.) The entropy of an entire closed system can never decrease within that system. Since the universe can be modeled as a closed system the universe is considered to be entropic – that is, running down.

The change in entropy (delta S) is equal to the heat transfer (delta Q) divided by the temperature (T).

$$\text{delta } S = (\text{delta } Q) / T$$

Second Law of Thermodynamics(heat engine): It is impossible to extract an amount of heat Q H from a hot reservoir and use it all to do work W. Some amount of heat Q C must be exhausted to a cold reservoir. This precludes a perfect heat engine. Second Law of Thermodynamics(refrigerator): It is not possible for heat to flow from a colder body to a warmer body without any work having been done to accomplish this flow. Energy will not flow spontaneously from a low temperature object to a higher temperature object. This precludes a perfect refrigerator

THE THIRD LAW OF THERMODYNAMICS

The Third Law of Thermodynamics is the lesser known of the three major thermodynamic laws. Together, these laws help form the foundations of modern science. The laws of thermodynamics are absolute physical laws – everything in the observable universe is subject to them. Like time or gravity, nothing in the universe is exempt from these laws. In its simplest form, the Third Law of Thermodynamics relates the entropy (randomness) of matter to its absolute temperature. The Third Law of Thermodynamics refers to a state known as "absolute zero." This is the bottom point on the Kelvin temperature scale. The Kelvin scale is absolute, meaning 0° Kelvin is mathematically the lowest possible temperature in the universe. This corresponds to about -273.15° Celsius, or -459.7 Fahrenheit.

In actuality, no object or system can have a temperature of zero Kelvin, because of the Second Law of Thermodynamics. The Second Law, in part, implies that heat can never spontaneously move from a colder body to a hotter body. So, as a system approaches absolute zero, it will eventually have to draw energy from whatever systems are nearby. If it draws energy, it can never obtain absolute zero. So, this state is not physically possible, but is a mathematical limit of the universe.

In its shortest form, the Third Law of Thermodynamics says: "The entropy of a pure perfect crystal is zero (0) at zero Kelvin (0° K)." Entropy is a property of matter and energy discussed by the Second Law of Thermodynamics. The Third Law of Thermodynamics means that as the temperature of a system approaches absolute zero, its entropy approaches a constant (for pure perfect crystals, this constant is zero). A pure perfect crystal is one in which every molecule is identical, and the molecular alignment is perfectly even throughout the substance. For non-pure crystals, or those with less-than perfect alignment, there will be some energy associated with the imperfections, so the entropy cannot become zero.

The Third Law of Thermodynamics can be visualized by thinking about water. Water in gas form has molecules that can move around very freely. Water vapor has very high entropy (randomness). As the gas cools, it becomes liquid. The liquid water molecules can still move around, but not as freely. They have lost some entropy. When the water cools further, it becomes solid ice. The solid water molecules can no longer move freely, but can only vibrate within the ice crystals. The entropy is now very low. As the water is cooled more, closer and closer to absolute zero, the vibration of the molecules diminishes. If the solid water reached absolute zero, all molecular motion would stop completely. At this point, the water would have no entropy (randomness) at all.

TEMPERATURE

Temperature is the property of a body or region of space that determines whether or not there will be a net flow of heat into it or out of it from a neighboring body or region and in which direction the heat will flow. If there is no heat flow the bodies or regions are said to be in thermal equilibrium and at the same temperature. If there is a flow of heat, the direction of the flow is from the body or region of higher temperature. Broadly, there are two methods of quantifying this property. The empirical method is to take two or more reproducible temperature-dependent events and assign fixed points on a scale of values to these events. For example, the Celsius temperature scale uses the freezing point and boiling point of water as the two fixed points, assigns the values 0 and 100 to them, respectively, and divides the scale between them into 100 degrees.

This method is serviceable for many practical purposes but lacking a theoretical basis it is awkward to use in many scientific contexts. In the 19th century, Lord Kelvin proposed a thermodynamic method to specify temperature, based on the measurement of the quantity of heat flowing between bodies at different temperatures. This concept relies on an absolute scale of temperature with an absolute zero of temperature, at which no body can give up heat. He also used Sadi Carnot's concept of an ideal frictionless perfectly efficient heat engine. This Carnot engine takes in a quantity of heat q1 at a temperature T1, and exhausts heat q2 at T2, so that T1/T2 = q1/q2. If T2 has a value fixed by definition, a Carnot engine can be run between this fixed temperature and any unknown temperature T1, enabling T1 to be calculated by measuring the values of q1 and q2. This concept remains the basis for defining thermodynamic temperature, quite independently of the nature of the working substance.

The unit in which thermodynamic temperature is expressed is the kelvin. In practice, thermodynamic temperatures cannot be measured directly; they are usually inferred from measurements with a gas thermometer containing a nearly ideal gas.

This is possible because another aspect of thermodynamic temperature is its relationship to the internal energy of a given amount of substance. This can be shown most simply in the case of an ideal monatomic gas, in which the internal energy per mole (U) is equal to the total kinetic energy of translation of the atoms in one mole of the gas (a monatomic gas has no rotational or vibrational energy). According to the kinetic theory, the thermodynamic temperature of such a gas is given by T = 2U/3R, where R is the universal gas constant.

THE KELVIN TEMPERATURE SCALE

Absolute Temperature

The Kelvin temperature scale (K) was developed by Lord Kelvin in the mid 1800s. The zero point of this scale is equivalent to -273.16 °C on the Celsius scale. This zero point is considered the lowest possible temperature of anything in the universe. Therefore, the Kelvin scale is also known as the "absolute temperature scale". At the freezing point of water, the temperature of the Kelvin scale reads 273 K. At the boiling point of water, it reads 373 K. Whereas the Kelvin scale is widely used by scientists, the Celsius or Fahrenheit scales are used in daily life. These two scales are easier to understand than the large numbers of the Kelvin scale. Could you imagine waking up to your radio and hearing the DJ give a weather report like this: "It's going to be a beautiful day today with sunny skies and a balmy temperature of 297 K!" That's 24 °C or 75 °F.

Pressure-Volume (Constant Temperature)

What happens to the volume of a gas as the pressure on it changes. Let's try the following experiment using equipment that might be found in your kitchen. Marshmallows are a mixture of sugar, air, and gelatin. The sugar makes them sweet, the air makes them fluffy, and the gelatin holds everything together. Marshmallows are a frozen foam and are mostly air by volume. When placed in a vacuum pump, they expand as the pressure decreases. Break the seal on their container and they shrink during the return to normal atmospheric pressure. Since the vacuum pump pulls on the marshmallows hard enough to burst some of the air bubbles, they are actually a bit smaller and more shriveled at the end of this experiment. The pressure of a gas is inversely proportional to its volume when temperature is constant. Symbolically .

$$P = 1/V$$

This correlation was discovered independently by Robert Boyle of Ireland in 1662 and Edme Mariotte of France in 1676. In Great Britain, America, Australia, the West Indies and other remnants of the British Empire it is called Boyle's law, while in Continental Europe and other places it is called Mariotte's law.

Mariotte added the important provision that temperature remain constant. Boyle neglected to mention it, but the data he used to derive his law were most likely collected during a period in which the temperature did not experience any significant change.

Since the gas needs to be in thermal equilibrium with its environment (or some other heat reservoir) to maintain an even temperature, the pressure-volume relationship normally applies only to "slow" processes. The marshmallow-vacuum experiment shown above is an example of a "slow" process. The pressure is reduced at a rate slow enough that heat from the environment is able to keep the jar and its contents at nearly room temperature. Such a transformation that takes place without a change in temperature is said to be isothermal.

Pumping a bicycle tire with a hand pump is an example of a "fast" process. The work done pushing the piston transforms into an increase in the internal energy (and thus an increase in the temperature) of the air molecules within the pump. People familiar with hand bicycle pumps will attest to the fact that they get hot after use. Likewise, when a gas is allowed to expanded into a region of reduced pressure it does work on its surroundings.

The energy to do this work comes from the internal energy of the gas and so the temperature of the gas drops. You can experience this yourself without the aid of any apparatus other than your mouth. Purse your lips so that your mouth has only a tiny opening to the outside and blow hard. The air rushing from your mouth will be quite cool despite coming from the core of your body, which is normally quite hot (around 37 °C). During a "fast" process like the ones just described, pressure and volume are changing so rapidly that heat doesn't have enough time to get into or out of the gas to keep the temperature constant. Such a transformation that takes place without any flow of heat is said to be adiabatic

$$P1\ V1 = P2\ V2 = \text{constant}$$

Pressure Temperature (Constant Volume)

The pressure of a gas is directly proportional to its temperature when volume is constant. Symbolically .

$$P \propto T$$

An isochoric process is one that takes place without any change in volume.

This relationship doesn't really have a name, but I have heard it called the "pressure law" or (mistakenly) "Gay-Lussac's law".

$$\frac{p_1}{T_1} = \frac{p_2}{T_2} = \text{constant}$$

IDEAL GASES

An ideal gas is a substance possessing very simple thermodynamic properties to which actual gases and vapours appear to approximate indefinitely at low pressures and high temperatures. It has the characteristic equation pv=Re, and obeys Boyle's law at all temperatures. The coefficient of expansion at constant pressure is equal to the coefficient of increase of pressure at constant volume. The difference of the specific heats by equation is constant and equal to R. The isothermal elasticity – v(dp/dv) is equal to the pressure p. The adiabatic elasticity is equal to y p, where-y is the ratio S/s of the specific heats. The heat absorbed in isothermal expansion from vo to v at a temperature 0 is equal to the work done by equation (since d0 = o, and 0(dp/d0)dv =pdv), and both are given by the expression RO log e (v/vo).

The energy E and the total heat F are functions of the temperature only and their variations take the form dE = sdO, d F = Sd0. The specific heats are independent of the pressure or density by equations. If we also assume that they are constant with respect to temperature (which does not necessarily follow from the characteristic equation, but is generally assumed, and appears from Regnault's experiments to be approximately the case for simple gases), the expressions for the change of energy or total heat from 00 to 0 may be written E – E0 = s(0 – 0 0), F – Fo = S(0-00). In thiscase the ratio of the specific heats is constant as well as the difference, and the adiabatic equation takes the simple form, pv v = constant, which is at once obtained by integrating the equation for the adiabatic elasticity, – v(dp/dv) =yp.

DEVIATIONS OF ACTUAL GASES FROM THE IDEAL STATE

Since no gas is ideally perfect, it is most important for practical purposes to discuss the deviations of actual gases from the ideal state, and to consider how their properties may be thermodynamically explained and defined. The most natural method of procedure is to observe the deviations from Boyle's law by measuring the changes of pv at various constant temperatures. It is found by experiment that the change of pv with pressure at moderate pressures is nearly proportional to the change of p, in other words that the coefficient d(pv)/dp is to a first approximation a function of the temperature only.

This coefficient is sometimes called the " angular coefficient," and may be regarded as a measure of the deviations from Boyle's law, 'which may be most simply expressed at moderate pressures by formulating the variation of the angular coefficient with temperature. But this procedure in itself is not sufficient, because, although it would be highly probable that a gas obeying Boyle's law at all temperatures was practically an ideal gas, it is evident that Boyle's law would be satisfied by any substance having the characteristic equation pv = f (0), where f (0) is any arbitrary function of 0, and that the scale of temperatures given by such a substance would not necessarily coincide with the absolute scale.

A sufficient test, in addition to Boyle's law, is the condition dE/dv=o at constant temperature. This gives by equation the condition Odp/d0 =p, which is satisfied by any substance possessing the characteristic equation p/0=f(v), where f(v) is any arbitrary function of v. This test was applied by Joule in the well-known experiment in which he allowed a gas to expand from one vessel to another in a calorimeter without doing external work.

Under this condition the increase of intrinsic energy would be equal to the heat absorbed, and would be indicated by fall of temperature of the calorimeter. Joule failed to observe any change of temperature in his apparatus, and was therefore justified in assuming that the increase of intrinsic energy of a gas in isothermal expansion was very small, and that the absorption of heat observed in a similar experiment in which the gas was allowed to do external work by expanding against the atmospheric pressure was equivalent to the external work done. But owing to the large thermal capacity of his calorimeter, the test, though sufficient for his immediate purpose, was not delicate enough to detect and measure the small deviations which actually exist.

JOULE–THOMSON EFFECT

In physics, the Joule–Thomson effect or Joule–Kelvin effect describes the temperature change of a gas or liquid when it is forced through a valve or porous plug while kept insulated so that no heat is exchanged with the environment.

This procedure is called a throttling process or Joule-Thomson process. At room temperature, all gases except hydrogen, helium and neon cool upon expansion by the Joule-Thomson process.

The effect is named for James Prescott Joule and William Thomson, 1st Baron Kelvin who discovered it in 1852 following earlier work by Joule on Joule expansion, in which a gas undergoes free expansion in a vacuum.

Description

The adiabatic (no heat exchanged) expansion of a gas may be carried out in a number of ways. The change in temperature experienced by the gas during expansion depends on the initial and final pressure, but also on the manner in which the expansion is carried out.

- If the expansion process is reversible, meaning that the gas is in thermodynamic equilibrium at all times, it is called an isentropic expansion. In this scenario, the gas does positive work during the expansion, and its temperature decreases.
- In a free expansion, on the other hand, the gas does no work and absorbs no heat, so the internal energy is conserved. Expanded in this manner, the temperature of an ideal gas would remain constant, but the temperature of a real gas may either increase or decrease, depending on the initial temperature and pressure.
- The method of expansion in which a gas or liquid at pressure P1 flows into a region of lower pressure P2 via a valve or porous plug under steady state conditions and without change in kinetic energy, is called the Joule–Thomson process. During this process, enthalpy remains unchanged.

Temperature change of either sign can occur during the Joule-Thomson process. Each real gas has a Joule–Thomson (Kelvin) inversion temperature above which expansion at constant enthalpy causes the temperature to rise, and below which such expansion causes cooling. This inversion temperature depends on pressure; for most gases at atmospheric pressure, the inversion temperature is above room temperature, so most gases can be cooled from room temperature by isenthalpic expansion.

Physical Mechanism

As a gas expands, the average distance between molecules grows. Because of intermolecular attractive forces expansion causes an increase in the potential energy of the gas. If no external work is extracted in the process and no heat is transferred, the total energy of the gas remains the same because of the conservation of energy. The increase in potential energy thus implies a decrease in kinetic energy and therefore in temperature.

A second mechanism has the opposite effect. During gas molecule collisions, kinetic energy is temporarily converted into potential energy. As the average intermolecular distance increases, there is a drop in the number of collisions per time unit, which causes a decrease in average potential energy. Again, total energy is conserved, so this leads to an increase in kinetic energy (temperature).

Below the Joule–Thomson inversion temperature, the former effect (work done internally against intermolecular attractive forces) dominates, and free expansion causes a decrease in temperature. Above the inversion temperature, gas molecules move faster and so collide more often, and the latter effect (reduced collisions causing a decrease in the average potential energy) dominates: Joule-Thomson expansion causes a temperature increase.

The Joule–Thomson (Kelvin) Coefficient

The rate of change of temperature T with respect to pressure P in a Joule–Thomson process (that is, at constant enthalpy H) is the Joule–Thomson (Kelvin) coefficient ìJT. This coefficient can be expressed in terms of the gas's volume V, its heat capacity at constant pressure Cp, and its coefficient of thermal expansion á as:

$$\mu_{JT} \equiv \left(\frac{\partial T}{\partial P}\right)_H = \frac{V}{C_P}(\alpha T - 1).$$

The value of μJT is typically expressed in °C/bar (SI units: K/Pa) and depends on the type of gas and on the temperature and pressure of the gas before expansion.

All real gases have an inversion point at which the value of μJT changes sign. The temperature of this point, the Joule–Thomson inversion temperature, depends on the pressure of the gas before expansion.

In a gas expansion the pressure decreases, so the sign of ∂P is always negative. With that in mind, the following table explains when the Joule–Thomson effect cools or warms a real gas:

If the gas temperature is	then µJT is	since ∂P is	thus∂P must be	so the gas cools
below the inversion temperature	positive	always negative	negative	
above the inversion temperature	negative	always negative	positive	warms

Helium and hydrogen are two gases whose Joule–Thomson inversion temperatures at a pressure of one atmosphere are very low (e.g., about 51 K (–222 °C) for helium). Thus, helium and hydrogen warm up when expanded at constant enthalpy at typical room temperatures. On the other hand nitrogen and oxygen, the two most abundant gases in air, have inversion temperatures of 621 K (348 °C) and 764 K (491 °C) respectively: these gases can be cooled from room temperature by the Joule–Thomson effect.

For an ideal gas, ìJT is always equal to zero: ideal gases neither warm nor cool upon being expanded at constant enthalpy.In practice, the Joule–Thomson effect is achieved by allowing the gas to expand through a throttling device (usually a valve) which must be very well insulated to prevent any heat transfer to or from the gas. No external work is extracted from the gas during the expansion. The effect is applied in the Linde technique as a standard process in the petrochemical industry, where the cooling effect is used to liquefy gases, and also in many cryogenic applications (e.g. for the production of liquid oxygen, nitrogen, and argon).

Only when the Joule–Thomson coefficient for the given gas at the given temperature is greater than zero can the gas be liquefied at that temperature by the Linde cycle. In other words, a gas must be below its inversion temperature to be liquefied by the Linde cycle. For this reason, simple Linde cycle liquefiers cannot normally be used to liquefy helium, hydrogen, or neon.

Proof that Enthalpy Remains Constant in a Joule-Thomson Process

In a Joule-Thomson process the enthalpy remains constant. To prove this, the first step is to compute the net work done by the gas that moves through the plug. Suppose that the gas has a volume of V1 in the region at pressure P1 and a volume of V2 when it appears in the region at pressure P2. Then the work done on the gas by the rest of the gas in region 1 is P1 V1. In region 2 the amount of work done by the gas is P2 V2. So, the total work done by the gas is

$$P_2V_2 - P_1V_1$$

The change in internal energy plus the work done by the gas is, by the first law of thermodynamics, the total amount of heat absorbed by the gas (here it is assumed that there is no change in kinetic energy). In the Joule-Thompson process the gas is kept insulated, so no heat is absorbed. This means that

$$E_2 - E_1 + P_2V_2 - P_1V_1 = 0$$

where E1 and E2 denote the internal energy of the gas in regions 1 and 2, respectively.

The above equation then implies that:

$$H_1 = H_2$$

where H1 and H2 denote the enthalpy of the gas in regions 1 and 2, respectively.

Derivation of the Joule-Thomson (Kelvin) coefficient

A derivation of the formula

$$m_{JT} \equiv \left(\frac{\partial T}{\partial P}\right)_H = \frac{V}{C_P}(\alpha T - 1)$$

for the Joule–Thomson (Kelvin) coefficient.

The partial derivative of T with respect to P at constant H can be computed by expressing the differential of the enthalpy dH in terms of dT and dP, and equating the resulting expression to zero and solving for the ratio of dT and dP. It follows from the fundamental thermodynamic relation that the differential of the enthalpy is given by:

$dH = TdS + VdP$ (here, S is the entropy of the gas).

Expressing dS in terms of dT and dP gives:

$$dH = T\left(\frac{\partial S}{\partial T}\right)_P dT + \left[V + T\left(\frac{\partial S}{\partial P}\right)_T\right]dP$$

Using

$$C_P = T\left(\frac{\partial S}{\partial T}\right)_P, \text{ we can write:}$$

$$dH = C_P dT + \left[V + T\left(\frac{\partial S}{\partial P}\right)_T\right]dP.$$

The remaining partial derivative of S can be expressed in terms of the coefficient of thermal expansion via a Maxwell relation as follows. From the fundamental thermodynamic relation, it follows that the differential of the Gibbs energy is given by:

$$dG = -SdT + VdP.$$

The symmetry of partial derivatives of G with respect to T and P implies that:

$$\left(\frac{\partial S}{\partial P}\right)_T = -\left(\frac{\partial V}{\partial T}\right)_P = -V\alpha$$

where α is the coefficient of thermal expansion. Using this relation, the differential of H can be expressed as

$$dH = C_p dT + V(1 - T\alpha)dP.$$

Equating dH to zero and solving for dT/dP then gives:

$$\left(\frac{\partial T}{\partial P}\right)_H = \frac{V}{C_P}(\alpha T - 1).$$

IDEAL GAS MODEL OF THERMODYNAMICS

The Kinetic Theory picture of a gas is often called the Ideal Gas Model. It ignores interactions between molecules, and the finite size of molecules. In fact, though, these only become important when the gas is very close to the temperature at which it become liquid, or under extremely high pressure. In this lecture, we will be analyzing the behavior of gases in the pressure and temperature range corresponding to heat engines, and in this range the Ideal Gas Model is an excellent approximation. Essentially, our program here is to learn how gases absorb heat and turn it into work, and vice versa. This heat-work interplay is called thermodynamics.

Julius Robert Mayer was the first to appreciate that there is an equivalence between heat and mechanical work. The tortuous path that led him to this conclusion is described in an earlier lecture, but once he was there, he realized that in fact the numerical equivalence—how many Joules in one calorie in present day terminology—could be figured out easily from the results of some measurements of gas specific heat by French scientists. The key was that they had measured specific heats both at constant volume and at constant pressure. Mayer realized that in the latter case, heating the gas necessarily increased its volume, and the gas therefore did work in pushing to expand its container. Having convinced himself that mechanical work and heat were equivalent, evidently the extra heat needed to raise the temperature of the gas at constant pressure was exactly the work the gas did on its container.

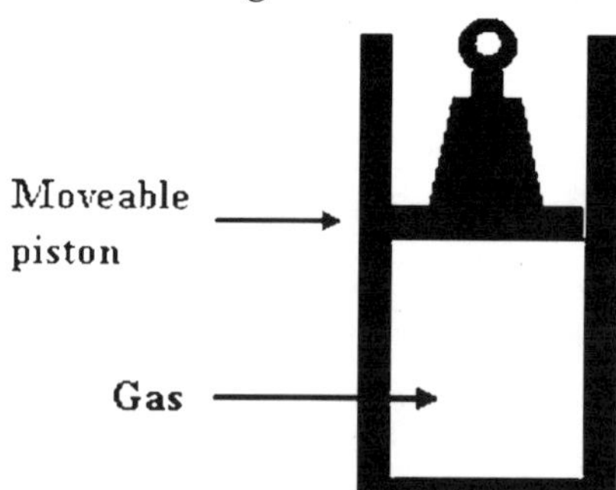

The simplest way to see what's going on is to imagine the gas in a cylinder, held in by a piston, carrying a fixed weight, able to move up and down the cylinder smoothly with negligible friction. The pressure on the gas is just the total weight pressing down divided by the area of the piston, and this total weight, of course, will not change as the piston moves slowly up or down: the gas is at constant pressure.

THE GAS SPECIFIC HEATS CV AND CP

Consider now the two specific heats of this same sample of gas, let's say one mole:

Specific heat at constant volume, C_V (piston glued in place),

Specific heat at constant pressure, C_P (piston free to rise, no friction).

In fact, we already worked out C_V in the Kinetic Theory lecture: at temperature T, recall the average kinetic energy per molecule is $\frac{3}{2}kT$, so one mole of gas—Avogadro's number of molecules—will have total kinetic energy, which we'll label internal energy,

$$E_{\text{int}} \frac{3}{2}kT.N_A \frac{3}{2}RT.$$

(In this simplest case, we are ignoring the possibility of the molecules having their own internal energy: they might be spinning or vibrating—we'll include that shortly).

That the internal energy is $\frac{3}{2}kT$ per mole immediately gives us the specific heat of a mole of gas in a fixed volume,

$$C_V = \frac{3}{2}R$$

that being the heat which must be supplied to raise the temperature by one degree.

However, if the gas, instead of being in a fixed box, is held in a cylinder at constant pressure, experiment confirms that more heat must be supplied to raise the gas temperature by one degree. As Mayer realized, the total heat energy that must be supplied to raise the temperature of the gas one degree at constant pressure is $\frac{3}{2}k$ per molecule plus the energy required to lift the weight. The work the gas must do to raise the weight is the force the gas exerts on the piston multiplied by the distance the piston moves. If the area of piston is A, then the gas at pressure P exerts force PA. If on heating through one degree the piston rises a distance Δh, the gas does work

$$PA.\Delta h = P\Delta V.$$

Now, for one mole of gas, $PV = RT$, so at constant P

$$P\Delta V = P\Delta T.$$

Therefore, the work done by the gas in raising the weight is just $R\Delta T$, the specific heat at constant pressure, the total heat energy needed to raise the temperature of one mole by one degree,

$$C_P = C_V + R.$$

In fact, this relationship is true whether or not the molecules have rotational or vibrational internal energy. (It's known as Mayer's relationship.) For example, the specific heat of oxygen at constant volume

$$C_V(\mathrm{O}_2) = \tfrac{5}{2}R.$$

and this is understood as a contribution of $\frac{3}{2}R$ from kinetic energy, and R from the two rotational modes of a dumbbell molecule (just why there is no contribution form rotation about the third axis can only be understood using quantum mechanics). The specific heat of oxygen at constant pressure

$$C_P(O_2) = \tfrac{3}{2}R.$$

It's worth having a standard symbol for the ratio of the specific heats:

$$\frac{C_P}{C_V} = \gamma.$$

ISOTHERMS AND ADIABATS

An ideal gas in a box has three thermodynamic variables: P, V, T. But if there is a fixed mass of gas, fixing two of these variables fixes the third from $PV = nRT$ (for n moles). In a heat engine, heat can enter the gas, then leave at a different stage. The gas can expand doing work, or contract as work is done on it. To track what's going on as a gas engine transfers heat to work, say, we must follow the varying state of the gas. We do that by tracing a curve in the (P, V) plane. Supplying heat to a gas which consequently expands and does mechanical work is the key to the heat engine. But just knowing that a gas is expanding and doing work is not enough information to follow its path in the (P, V) plane. The route it follows will depend on whether or not heat is being supplied (or taken away) at the same time. There are, however, two particular ways a gas can expand reversibly—meaning that a tiny change in the external conditions would be sufficient for the gas to retrace its path in the (P, V) plane backwards. It's important to concentrate on reversible paths, because as Carnot proved and we shall discuss later, they correspond to the most efficient engines. The two sets of reversible paths are the isotherms and the adiabats.

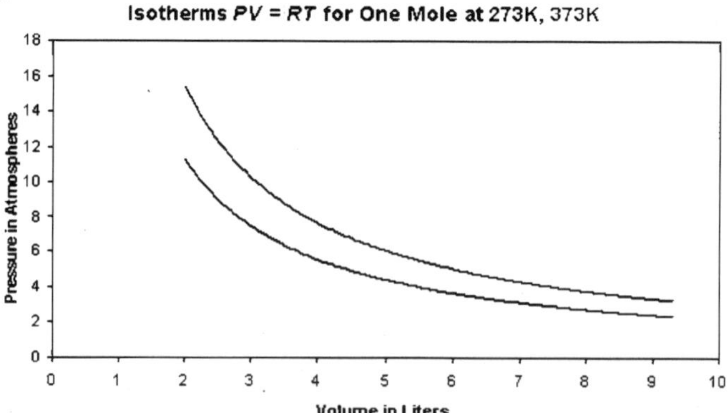

Isothermal behavior: The gas is kept at constant temperature by allowing heat flow back and forth with a very large object (a "heat reservoir") at temperature T. From $PV = nRT$, it is evident that for a fixed mass of gas, held at constant T but subject to (slowly) varying pressure, the variables P, V will trace a hyperbolic path in the (P, V) plane. This path, $PV = nRT_1$, say is called the isotherm at temperature T1. Here are two examples of isotherms:

Adiabatic behavior: "adiabatic" means "nothing gets through", in this case no heat gets in or out of the gas through the walls. So all the work done in compressing the gas has to go into the internal energy E_{int}.

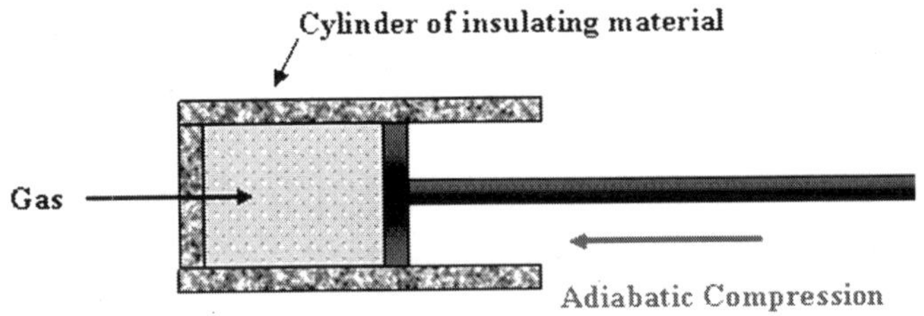

As the gas is compressed, it follows a curve in the (P, V) plane called an adiabat. To see how an adiabat differs from an isotherm, imagine beginning at some point on the blue 273K isotherm on the above graph, and applying pressure so the gas moves to higher pressure and lower volume.

Since the gas's internal energy is increasing, but the number of molecules is staying the same, its temperature is necessarily rising, it will move towards the red curve, then above it.

This means the adiabats are always steeper than the isotherms.

EQUATION FOR AN ADIABAT

What equation for an adiabat corresponds to $PV = nRT_1$ for an isotherm?

On raising the gas temperature by ΔT, the change in the internal energy—the sum of molecular kinetic energy, rotational energy and vibrational energy,

$$\Delta E_{int} = C_V \Delta T.$$

This is always true: whether or not the gas is changing volume is irrelevant, all that counts in E_{int} is the sum of the energies of the individual molecules (assuming as we do here that attractive or repulsive forces between molecules are negligible). In adiabatic compression, all the work done by the external pressure goes into this internal energy, so

$$-P\Delta V = C_V \Delta T.$$

(Compressing the gas of course gives negative ΔV, positive ΔEint.)

To find the equation of an adiabat, we take the infinitesimal limit

$$-PdV = C_V dT.$$

Divide the left-hand side by PV, the right-hand side by RT (since $PV = RT$, that's OK) to find

$$-\frac{R}{C_V}\frac{dV}{V} = \frac{dT}{T}.$$

Recall now that $C_P = C_V + R$, and $C_P/C_V = \gamma$. It follows that

$$\frac{R}{C_V} = \frac{C_P - C_V}{C_V} = \gamma - 1.$$

Hence

$$-(\gamma - 1)\int \frac{dV}{V} = \int \frac{dT}{T}$$

and integrating

$$\ln T + (\gamma - 1)\ln V = \text{const.}$$

from which the equation of an adiabat is

$$TV^{-1} = \text{const.}$$

From $PV - RT$, the P, V equation for an adiabat can be found by multiplying the left-hand side of this equation by the constant PV/T, giving

$$PV^{\gamma} = \text{const. for an adiabat,}$$

where $\gamma = {}^5/_3$ for a monatomic gas, ${}^7/_5$ for a diatomic gas.

THE CARNOT CYCLE

All standard heat engines (steam, gasoline, diesel) work by supplying heat to a gas, the gas then expands in a cylinder and pushes a piston to do its work. The catch is that the heat and/or the gas must somehow then be dumped out of the cylinder to get ready for the next cycle.

We examine the first step, the expansion, then go on to the full cycle—Carnot's analysis. Carnot's aim was to figure out how to maximize the efficiency of a heat engine, and then work out what that efficiency was, that is, how much of the heat supplied was actually converted into the mechanical work done by the engine.

Remember that he had in mind the analogy of the water wheel, at that time still a main driving force of industry. He knew that the most efficient water wheels were those that operated smoothly, the water went into the buckets at the top from the same level, it didn't fall through any height, and didn't splash around. In the limit of a frictionless wheel, with gentle flow on and off the wheel, the machine would be reversible—turning it in reverse to raise the water back would take the same amount of work the wheel had delivered as the water fell.

This was clearly perfect efficiency, so these were to conditions to emulate in the heat engine. The analog to having the water flow into buckets at the same height, with no wasteful drop, is to have the heat from the heat supply flow into the gas at the same temperature. There must of course be a slight drop in temperature for the heat to flow at all, but this must be minimized. This means that as the heat is supplied and the gas expands, the temperature of the gas stays the same as that of the heat supply (the "heat reservoir") and the gas is expanding isothermally.

ISOTHERMAL EXPANSION

So the first question is: how much work is done by an isothermally expanding gas? Taking the temperature of the heat reservoir to be TH (H for hot), the expanding gas follows the isothermal path $PV = nRT_H$ in the (P, V) plane.

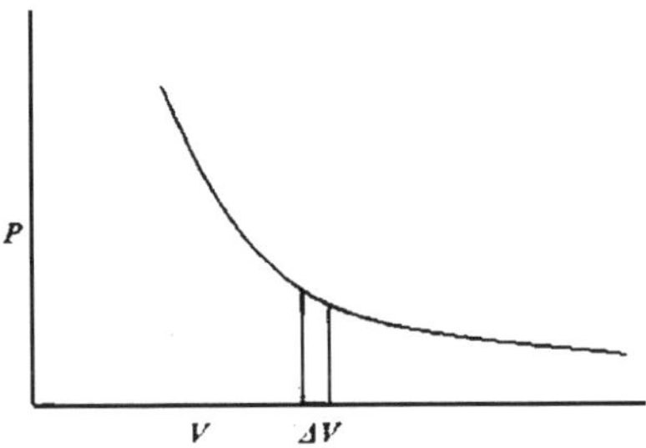

The work done by the gas in a small volume expansion ΔV is just $P\Delta V$, the area under the curve.Hence the work done in expanding isothermally from volume Va to Vb is the total area under the curve between those values,

$$\text{work done isothermally} = \int_{V_a}^{V_b} PdV = \int_{V_a}^{V_b} \frac{nRV_H}{V} dV = nRT_H \ln \frac{V_b}{V_a}.$$

Since the gas is at constant temperature TH, there is no change in its internal energy during this expansion, so the total heat supplied must be

$$nRT_H \ln \frac{V_b}{V_a},$$

the same as the external work the gas has done.

In fact, this isothermal expansion is only the first step: the gas is at the temperature of the heat reservoir, hotter than its other surroundings, and will be able to continue expanding even if the heat supply is cut off. To ensure that this further expansion is also reversible, the gas must not be losing heat to the surroundings. That is, after the heat supply is cut off, there must be no further heat exchange with the surroundings, the expansion must be adiabatic.

ADIABATIC EXPANSION

The work done in an adiabatic expansion is like that done in allowing a compressed spring to expand against a force—equal to the work needed to compress the spring in the first place, for a perfect spring, and an adiabatically enclosed gas is essentially perfect in this respect. In other words, adiabatic expansion is reversible. To find the work the gas does in expanding adiabatically from Vb to Vc, say, the above analysis is repeated with the isotherm $PV = nRT_H$ replaced by the adiabat $PV^\gamma = P_b V_b^\gamma$, work done adiabatically W_{adiabat}

$$\int_{V_a}^{V_c} PdV = P_b V_b^\gamma \int_{V_b}^{V_c} \frac{dV}{V^\gamma} = P_b V_b^\gamma \frac{V_c^{1-\gamma} - V_b^{1-\gamma}}{1-\gamma}.$$

Again, this is the area under the curve, in this case under the adiabat, from b to c in the (P, V) plane.

Since points b, c are on the same adiabat, $P_b V_b^{\gamma} = P_c V_c^{\gamma}$, and the expression can be written more neatly:

$$W_{\text{adiabat}} = \frac{P_c V_c - P_b V_b}{1-\gamma}.$$

This is a useful expression for the work done since we are plotting in the (P, V) plane, but note that from the gas law $PV = nRT$, the numerator is just $nR(Tc - T_b)$, and from this $W_{\text{adiabat}} = nC_V(T_c - T_b)$, as of course it must be—this is the loss of internal energy that has been expended by the gas on expanding against external pressure.

We've looked in detail at the work a gas does in expanding as heat is supplied (isothermally) and when there is no heat exchange (adiabatically). These are the two initial steps in a heat engine, but it is equally necessary for the engine to get back to where it began, for the next cycle. The general idea is that the piston drives a wheel, which continues to turn and pushes the gas back to the original volume.

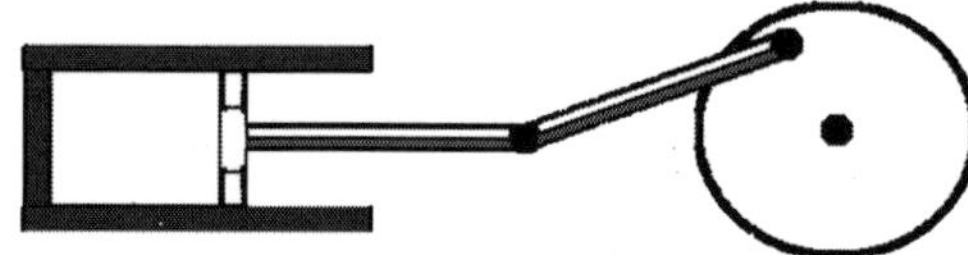

But it is also essential for the gas to be as cold as possible on this return leg, because the wheel is now having to expend work on the gas, and we want that to be as little work as possible—it's costing us. The colder the gas, the less pressure the wheel is pushing against.

To ensure that the engine is as efficient as possible, this return path to the starting point (P_a, V_a)must also be reversible. We can't just retrace the path taken in the first two legs, that would take all the work the engine did along those legs, and leave us with no net output. Now the gas cooled during the adiabatic expansion from b to c, from TH to TC, say, so we can go some distance back along the reversible colder isotherm TC. But this won't get us back to (P_a, V_a), because that's on the TH isotherm. The simplest option—the one chosen by Carnot—is to proceed back along the cold isotherm to the point where it intersects the adiabat through a, then follow that isotherm back to a. (One could follow a more complicated path: provided it was composed of segments each being adiabatic or isothermal, it would be reversible.) Carnot's cycle is around that curved quadrilateral having these four curves as its sides.

Efficiency of the Carnot Engine

In a complete cycle of Carnot's heat engine, the gas traces the path abcd. The important question is: what fraction of the heat supplied from the hot reservoir (along the red top isotherm) is turned into mechanical work? This fraction is called the efficiency of the engine. The work output along any curve in the (P, V) plane is just $\int PdV$ —the area under the curve, but it will be negative if the volume is decreasing! So the work done by the engine during the hot isothermal segment is the area abfh, then the adiabatic expansion adds the area bcef, but as the gas is compressed back, the wheel has to do work on the gas equal to the area cdge as heat is dumped into the cold reservoir, then dahg as the gas is recompressed to the starting point.

The bottom line is that the total work done by the gas is the area bounded by the four paths: the curved "parallelogram" in the picture above. We could compute this area by finding $\int PdV$ for each segment, but that is unnecessary— on completing the cycle, the gas is back to its initial temperature, so has the same internal energy. Therefore, the work done by the engine must be just the difference between the heat supplied at TH and that dumped at TC.

Now the heat supplies along the initial hot isothermal path ab, equal to the work done along that leg, is (from the paragraph above on isothermal expansion):

$$Q_H = nRT_H \ln \frac{V_b}{V_a}$$

and the heat dumped into the cold reservoir along cd is

$$Q_C = nRT_C \ln \frac{V_c}{V_d}.$$

The difference between these two is the net work output. This can be simplified using the adiabatic equations for the other two sides of the cycle:

$$T_H V_b^{\gamma-1} = T_C V_c^{\gamma-1}$$

$$T_H V_a^{\gamma-1} = T_C V_d^{\gamma-1}.$$

Dividing the first of these equations by the second,

$$\left(\frac{V_b}{V_a}\right) = \left(\frac{V_c}{V_d}\right)$$

and using that in the preceding equation for QC,

$$Q_C = nRT_C \ln \frac{V_a}{V_b} = \frac{T_C}{T_H} Q_H.$$

The work done can now be written simply:

$$W = Q_H - Q_C = \left(1 - \frac{T_C}{T_H}\right) Q_H.$$

Therefore the efficiency of the engine, defined as the fraction of the ingoing heat energy that is converted to available work, is

$$\text{efficiency} = \frac{W}{Q_H} = 1 - \frac{T_C}{T_H}.$$

These temperatures are of course in degrees Kelvin, so for example the efficiency of a Carnot engine having a hot reservoir of boiling water and a cold reservoir ice cold water will be 1– (273/373) = 0.27, just over a quarter of the heat energy is transformed into useful work.After all the effort to construct an efficient heat engine, making it reversible to eliminate "friction" losses, etc., it is perhaps somewhat disappointing to find this figure of 27% efficiency when operating between 0 and 100 degrees Celsius. Surely we can do better than that? After all, the heat energy of hot water is the kinetic energy of the moving molecules, can't we find some device to channel all that energy into useful work? Well, we can do better than 27%, by having a colder cold reservoir, or a hotter hot one. But there's a limit: we can never reach 100% efficiency, because we cannot have a cold reservoir at $T_C = 0K$, and even if we did after the first cycle the heat dumped into it would warm it up!

The Second Law of Thermodynamics states that we cannot devise an engine, working in a cycle, that simply extracts heat from a hot reservoir and delivers mechanical work.

This means any engine that takes heat and delivers work also dumps out some of the initial heat to a reservoir at a lower temperature. It's important to note that the First Law of Thermodynamics, the conservation of total energy including heat, would not be violated by an engine that powered a ship by extracting heat energy from the surrounding water.

This Second Law is saying something new. And, this Second Law does not follow from the First by logical deduction—it comes (like the First) from experiment and observation.

THE MOST EFFICIENT ENGINE

An important consequence of the second Law is that no engine can be more efficient than the Carnot cycle. Essentially, this is because a "super efficient" engine, if one existed, could be used to drive a Carnot cycle in reverse, which would pump back to the hot reservoir the heat the super efficient engine dumped in the cold reservoir, and the net effect of the two coupled engines would be to take heat from the hot reservoir and do work, contradicting the Second Law.

To see this, we plot the heat/energy flow for the Carnot cycle:

Here $Q_H = Q_C + W$ (all expressed in Joules, of course).

Since the engine is reversible, it can also be run backwards (this would be a refrigerator: outside work is supplied, and heat is extracted from a cold reservoir and dumped into a hot reservoir:

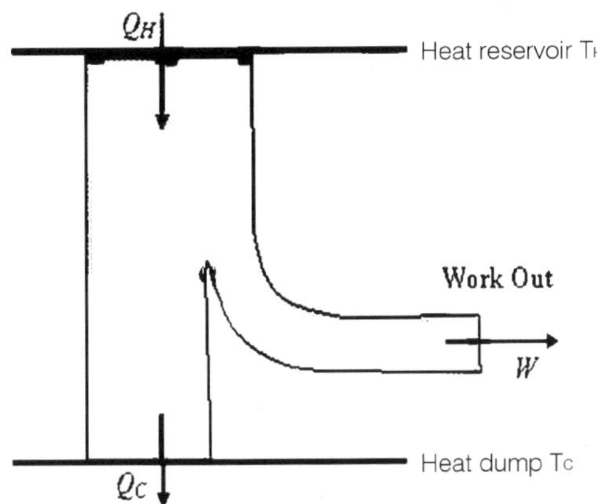

Suppose now we have a super efficient engine, represented by the first diagram above, and dumping the same heat per cycle QC into the cold reservoir, but taking in more heat energy $Q_H + \Delta$ Joules from the hot reservoir, and performing work $W + \Delta$. Now, we hook up our super efficient engine to the "Carnot refrigerator" in the diagram above. The refrigerator sucks out of the cold reservoir all the heat the super efficient engine dumped there, and needs W Joules of work per cycle to do it.

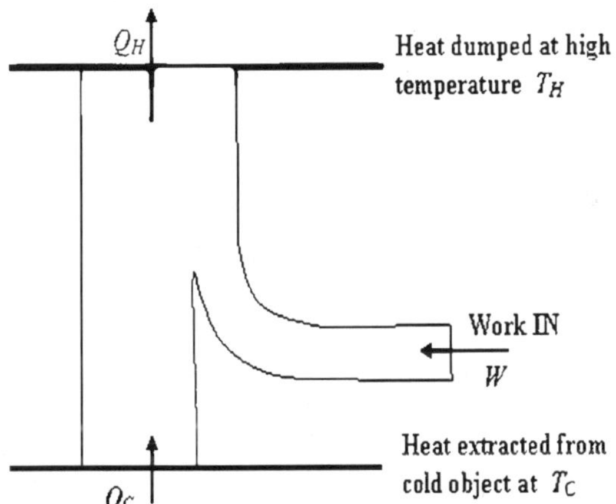

The super efficient engine can provide this, and there are still Δ Joules of work to spare. Of course, the Carnot refrigerator has also dumped QH Joules of heat in the hot reservoir. But the bottom line is that between them, the super efficient engine and the Carnot refrigerator have extracted Δ Joules of heat from the hot reservoir and performed Δ Joules of work—contradicting the Second Law.

The Second Law therefore forces the conclusion that no amount of machine design will produce an engine more efficient than the Carnot cycle. The rather low ultimate efficiencies this dictated came as a shock to nineteenth century engineers.

MOLECULAR COLLISIONS

In analyzing the gas so far, we've ignored collisions between molecules, and in fact for air at ordinary temperatures the relationship between pressure, volume and temperature came out correctly. Furthermore, Maxwell's speed distribution can be used to find what fraction of the molecules in a planet's atmosphere are moving at above escape velocity, so we can predict what gases will remain surrounding a planet, given the gravitational force near the surface, and the temperature. But there are other phenomena for which an understanding of collisions is all-important.

For example, if two different gases, say oxygen and nitrogen, in a container at the same pressure and temperature are separated by a partition, how quickly will they mix once the partition is removed? Assuming room temperature, the molecules will be moving at hundreds of meters per second, so one might imagine the mixing will be over in hundredths of a second. But that is not the case at all—observationally, it might take an hour, for a box holding a few liters. This surprisingly slow penetration of one gas by another is called diffusion. The reason it takes the gases so long to mix becomes evident on tracking one molecule as it enters the other gas. Think of an oxygen molecule moving into nitrogen. We'll take O_2 and N_2 to be little spheres of diameter d. So now visualize the little O_2 sphere shooting into this space where all these other spheres are moving around. Temporarily, for ease of visualization, let's imagine all the other spheres to be at rest.

How far can we expect the O_2 to get before it hits an N_2? The average distance before a collision is called the mean free path. Let's try to picture how much room there is to fly between these fixed N_2 spheres. We do know that if it were liquid nitrogen, there would be very little room: liquids are just about incompressible, so the molecules must be touching. Roughly speaking, a molecule of diameter d will occupy a cubical volume of about d3 (there has to be some space left over—we can pack cubes to fill space, but not spheres).

We also know that liquid nitrogen weighs about 800 kg per cubic meter, whereas N_2 gas at room temperature (and pressure) weighs about 1.2 kg per cubic meter, a ratio of 670. This means that on average each molecule in the gas has 670 times more room—that is, it has a space 670 times the volume d3 we gave it in the liquid. So in the gas, the average center-to-center separation of the molecules will be the cube root of 670, which is about 8.75d. So the picture is a gas of spheres of diameter d, placed at random, but separated on average by distances of order 10d. It's clear that shooting an oxygen molecule into this it will get quite a way.

We now try to estimate just how far an O_2 will get, on average, as it shoots into this forest of spheres. Picture the motion of the center of the oxygen molecule. Before any collision, it will be moving on a straight-line path. Just how close does the O_2 center have to get to an N2 center for a hit? Taking both O_2, N_2 to be spheres of diameter d, if an N2 center lies within d of the O_2 center's path, there will be a hit. So we can think of the O_2 as sweeping out a volume, a cylinder of radius d centered on its path, hitting and deflecting if it encounters an N_2 centered within that cylinder. So how far will it get, on average, before a hit? In traveling a distance x, it sweeps out a volume $\pi d^2 x$. Now picture it going through the gas for some considerable length of time, so there are many collisions.

The volume swept out will look like a stovepipe, long straight cylindrical sections connected by elbows at the collisions. The total volume of this stovepipe (ignoring tiny corrections from the elbows) will be just $\pi d^2 L$, L being the total length, that is, the total distance the molecule traveled. If the density of the nitrogen is n molecules per cubic meter, the number of N2's in this stovepipe volume will be $\pi d^2 Ln$, in other words, this will be the number of collisions. Therefore, the average distance between collisions, the mean free path l, is given by:

$$\text{mean free path } l = \frac{\text{total distance traveled}}{\text{number of collisions}} = \frac{L}{\pi d^2 Ln} = \frac{L}{\pi d^2 n}.$$

So what is n? We estimated above that each molecule has space 670d 3 to itself, so n is just how many of those volumes there are in one cubic meter, that is, $n = 1/670d^3$.

Therefore, the mean free path is given by

$$l = \frac{1}{\pi d^2 n} = \frac{670d^3}{\pi d^3} = 200d.$$

Notice that this derivation of the mean free path in terms of the molecular diameter depends only on knowing the ratio of the gas density to the liquid density—it does not depend on the actual size of the molecules! But it does mean that if we can somehow measure the mean free path, by measuring how fast one gas diffuses into another, for example, we can deduce the size of the molecules, and historically this was one of the first ways the size of molecules was determined, and so Avogadro's number was found.

Let us now put in some numbers to find this mean free path: for O_2, N2, d = 0.3 nm, so the mean free path l = 60nm, or 6×10^{-8}m. The speed of the molecules at room temperature v is approximately 500 meters per sec., so the molecule has of order 1010 collisions per second!

Actually, there is one further correction we should make. We took the N2 molecules to be at rest, whereas in fact they're moving as fast as the oxygen molecule, approximately. This means that even if the O_2 is temporarily at rest, it can undergo a collision as an N2 comes towards it. Clearly, what really counts in the collision rate is the relative velocity of the molecules. Defining the average velocity as the root mean square velocity, if the O_2 has velocity $\vec{v}_1$ and the N2 $\vec{v}_2$, then the square of the relative velocity $\overline{(\vec{v}_1 - \vec{v}_2)^2} = \overline{\vec{v}_1^2} -$

$\overline{2\vec{v}_1 - \vec{v}_2}^2 + \overline{\vec{v}_2^2} = \overline{\vec{v}_1^2} + \overline{\vec{v}_2^2}$, since $\overline{\vec{v}_1.\vec{v}_2}$ must average to zero, the relative directions being random. So the average square of the relative velocity is twice the average square of the velocity, and therefore the average root-mean-square velocity is up by a factor –2, and the collision rate is increased by this factor. Consequently, the mean free path is decreased by a factor of –2 when we take into account that all the molecules are moving.

Our final result, then, is that the mean free path

$$1 = \frac{1}{\sqrt{2}\pi d^2 n}.$$

Finding the mean free path is—literally—the first step in figuring out how rapidly the oxygen atoms will diffuse into the nitrogen gas, and of course vice versa.

What we really want to know is just how much we can expect the gases to have intermingled after a given period of time. We'll just follow the one molecule, and estimate how far it gets. To begin with, let's assume for simplicity that it tales steps all of the same length l, but after each collision it bounces off in a random direction. So after N steps, it will have moved to a point

$$\vec{L} = \vec{l}_1 + \vec{l}_2 + \vec{l}_3 + \cdots + \vec{l}_N,$$

where each vector $\vec{l}_j$ has length l, but the vectors all point in random different directions.

If we now imagine many of the oxygen molecules following random paths like this, how far on average can we expect them to have drifted after N steps? (note that they could with equal likelihood be going backwards!) The appropriate measure is the root-mean-square distance,

$$\overline{\vec{L}^2} = \overline{\left(\vec{l}_1 + \vec{l}_2 + \vec{l}_3 \ldots \vec{l}_N\right)^2} = Nl^2 + \sum_{i.j} \overline{\vec{l}_i.\vec{l}_j}$$

Since the direction after each collision is completely random, $\overline{\vec{l}_i.\vec{l}_j} = 0$, and the root-mean-square distance $\sqrt{\overline{\vec{L}}^2} = \sqrt{N}l.$

If we allow steps of different lengths, the same argument works, but now l is the root-mean-square path length.

The important factor here is the $\sqrt{N}$. Recall from above that $l = 60$ nm, or 6×10^{-8} m., and there are of order 1010 collisions per second. This means that the average distance diffused in one second is $\sqrt{10^{10}}l = 10^5 l$, say half a centimeter. The average distance in one hour would be only 60 times this, or 30 cm., one foot, and in a day about five feet—the average distance traveled is only increasing as the square root of the time elapsed!

This is a very general result. For example, suppose we have a gas in which the mean free path is l and the average speed of the molecules is v. Then the average time between collisions $\tau = l/v$. The number of collisions in time t will be t/τ, so the average distance a molecule moves in time t will be $r = l\sqrt{t/\tau}$.

A famous mystery cleared up by arguments like this was that Newton predicted the speed of sound would be given by $c^2 = B/\rho$, as we discussed earlier in the course, with B the bulk modulus. But when B was measured carefully by slowly compressing air, the result was in error by about 30%! The speed of sound predicted a higher (stiffer) bulk modulus.

The explanation turned out to be that in a slow measurement of the bulk modulus, the gas stays at the same temperature—the heating caused by slow compression leaks away. But if the compression is rapid, the gas heats up and so the pressure goes up more than if it had stayed at the same temperature. So the question is whether the compression and decompression as a sound wave passes through is so rapid that the heated-up gas doesn't have time to spread to the cooled regions.

For sound at say 1000Hz, the wavelength is 34 cm. If compression heats gas locally, the hot molecules will diffuse away in a similar manner to that discussed above. They will be slightly faster than the average molecules. In 1/1000 th of a second, they will have 107 collisions, so will travel about $\sqrt{10^7}\ l$ = 30C 0l = 0.2mm. This tiny distance compared with the wavelength of the sound wave means that during the compression/decompression cycles as the wave passes through, the heat has no chance to dissipate—so, effectively, it's like compressing a gas in an insulated container, it's harder to compress than it would be if the heat generated could flow away, and the bulk modulus is higher by an amount (around 30%) we shall work out in a forthcoming lecture.

The simple kinetic picture of a gas leads to the Gas Law, $PV = nRT$, with the macroscopic variable T now seen as proportional to the average energy of a molecule. Putting heat into a gas simply means increasing the molecules' average kinetic energy (plus rotational energy, etc., for more complicated molecules) and the work done by an expanding gas is just a transfer of energy from this internal energy to whatever the gas is pushing.

The First Law of Thermodynamics is nothing but conservation of total energy, now that we know heat is really microscopic kinetic energy for an ideal gas (and rotational energy, etc.). By applying these findings to the Carnot Cycle, we discovered that this reversible ideal engine had efficiency $1 - T_C/T_H$. Furthermore, we found that no heat engine could be more efficient, at least if we granted the truth of the Second Law of Thermodynamics, which states that it is impossible to construct an engine that just takes heat energy out of the air, say, and converts it into work, without any available "cold reservoir" to dump waste heat into.

The incredible thing is that Carnot, in the 1820's, who believed heat was a conserved caloric fluid, and therefore didn't know the First Law (that heat is just microscopic energy, which is not conserved by itself, but can be transformed into other forms of energy) correctly found the efficiency of his cycle to be $1 - T_C/T_H$, and then argued—also correctly—that this set the absolute limit on heat engine efficiency! How did he do that? It's worth a brief examination of his ideas, because they give a clue about something we haven't mentioned so far—entropy.

Remember, Carnot saw his cycle as a water wheel, the caloric fluid being the water, the temperature difference being the height difference from the top to the bottom of the waterfall. But he did know one difference from a waterfall: he knew there was an absolute zero, a level below which nothing could "fall". (This was from the well known results on the contraction of a gas with cooling. He took the absolute zero to be -267 Celsius, only a few degrees off the correct value). This meant that one could assign to the caloric fluid an absolute value of "potential energy" in this imaginary water wheel scenario: at temperature T, an amount of fluid F would have "potential energy" FT. This is the total possible amount of heat energy in the fluid, and to extract it all you would have to "drop" the fluid all the way down the temperature scale to the absolute zero. But in our real (if idealized) heat engine, it can only drop as far as the lowest temperature available, that of the cold reservoir TC. Hence the efficiency factor $1-T_C/T_H$: it's just the ratio of how far the caloric fluid is able to fall in our engine to how far it would have to fall (down from TH to absolute zero) to give up all its heat energy.

In this picture, the heat energy delivered from the hot reservoir $Q_H = FT_H$, so the amount of actual "caloric fluid" must be $F = Q_H/T_H$. But the fluid is conserved, in Carnot's picture, so the waste heat is $Q_C = FT_C$, with the same F, and therefore $Q_H/T_H = Q_C/T_C$. This equation is the important one: even though the argument he used is incorrect—there is no conserved caloric fluid—the equation is right, and relates directly to the efficiency of his engine: the heat utilized is $Q_H - Q_C = Q_H (1- T_C/T_H)$. But how did he know no engine could do better? Ironically, although he didn't know the First Law of Thermodynamics, he had figured out the Second Law, and he presented an argument exactly like that given in the last lecture to demonstrate that no engine could outperform his reversible one.

Let's go back to the Carnot cycle and consider a bit more how the gas gains heat from a reservoir as it follows a path in the (P, V) plane. We begin with the first half of the cycle, from a to c:

ISOTHERM

This does not mean that we can say the gas at (P_C, V_C) has QH more heat than the gas at (P_a, V_a). Why not? Because we could equally well have gone from a to c by a route which is the second half of the Carnot cycle traveled backwards:

This is a perfectly well defined reversible route, ending at the same place, but with quite a different amount of heat supplied. So we cannot say that a gas at a given (P, V) contains a definite amount of heat. It does of course have a definite internal energy, but that energy can be increased by adding a mix of external work and supplied heat, and the two different routes from a to c have the same total energy supplied to the gas, but with more heat and less work along the top route.

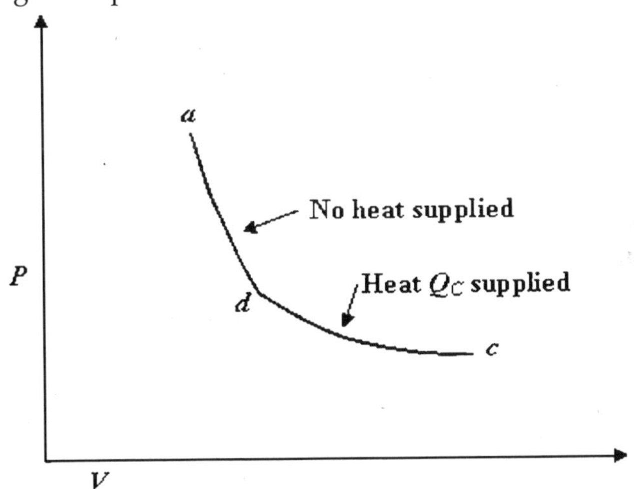

However, notice that one thing is the same over the two routes in the diagrams above: the ratio of the heat supplied to the temperature at which is was delivered: $Q_H/T_H = Q_C/T_C$.

Of course, we've chosen two particular reversible routes from a to c, but it turns out that for any reversible route from a to c, the integral of increments of heat supplied divided by the temperature of delivery, $\int_a^c dQ/T$, is the same! To see how this can be, consider cutting a corner in the previous route:

If we follow the path aefgc instead of adc, where ef is an isotherm and fg an adiabat, how does that affect $\int_a^c dQ/T$? The answer is it doesn't: look at the little Carnot cycle efgd. We've just changed from the bottom route to the top route around this cycle from e to g, so from the previous argument about the original big Carnot cycle, $\int_e^g dQ/T$ is the same.

But we can now cut corners on the corners: any zigzag route from a to c, with the zigs isotherms and the zags adiabats, in other words, any reversible route, can be constructed by adding little Carnot cycles to the original route. In fact, any path you can draw in the plane from a to c can be approximated arbitrarily well by a reversible route made up of little bits of isotherms and adiabats.

The value of $\int_a^c dQ/T$ is evidently the same along any reversible path from a to c: in contrast to $\int_a^c dQ$, the heat change, which was different for different reversible routes.

We can define a new state variable, S, such that the difference

$$S(P_2,V_2) = S(P_1,V_1) + \int_{(P_1,V_1)}^{(P_2,V_2)} \frac{dQ}{T}$$

where the integral is understood to be along a reversible path. Given the value of S at a single point (P_1,V_1), we can find its value everywhere.

2

Boltzmann Distributions and Thermodynamics

BOLTZMANN DISTRIBUTIONS

We have gained some understanding of the macroscopic properties of the air around us. For instance, we know something about its internal energy and specific heat capacity. How can we obtain some information about the statistical properties of the molecules which make up air? Consider a specific molecule: it constantly collides with its immediate neighbour molecules, and occasionally bounces off the walls of the room. These interactions "inform" it about the macroscopic state of the air, such as its temperature, pressure, and volume. The statistical distribution of the molecule over its own particular microstates must be consistent with this macrostate. In other words, if we have a large group of such molecules with similar statistical distributions, then they must be equivalent to air with the appropriate macroscopic properties. So, it ought to be possible to calculate the probability distribution of the molecule over its microstates from a knowledge of these macroscopic properties.

We can think of the interaction of a molecule with the air in a classroom as analogous to the interaction of a small system A in thermal contact with a heat reservoir A¢. The air acts like a heat reservoir because its energy fluctuations due to any interactions with the molecule are far too small to affect any of its macroscopic parameters. Let us determine the probability P_r of finding system A in one particular microstate $r$of energy E_r when it is thermal equilibrium with the heat reservoir A'.

As usual, we assume fairly weak interaction between A and A', so that the energies of these two systems are additive. The energy of A is not known at this stage. In fact, only the total energy of the combined system $A^{(0)} = A + A'$ is known. Suppose that the total energy lies in the range $E^{(0)}$ to $E^{(0)} + \mathrm{d}E$. The overall energy is constant in time, since $A^{(0)}$ is assumed to be an isolated system, so

$$E_r + E' = E^{(0)},$$

where E' denotes the energy of the reservoir A'. Let $\Omega'(E')$ be the number of microstates accessible to the reservoir when its energy lies in the range E' to $E' + \delta E$ Clearly, if system A has an energy E_r then the reservoir A¢ must have an energy close to $E' = E^{(0)} - E_r$. Hence, since A is in one definite state (i.e., state r), and the total number of states accessible to A' is $\Omega'(E^{(0)} - E_r)$ it follows that the total number of states accessible to the combined system is simply $\Omega'(E^{(0)} - E_r)$. The principle of equal a priori probabilities tells us the the probability of occurrence of a particular situation is proportional to the number of accessible microstates. Thus,

$$P_r = C'\,\Omega'\,(E^{(0)} - E_r),$$

where C' is a constant of proportionality which is independent of r. This constant can be determined by the normalization condition

$$\sum_r P_r = 1,$$

where the sum is over all possible states of system A, irrespective of their energy.

Let us now make use of the fact that system A is far smaller than system A'. It follows that $E_r << E^{(0)}$, so the slowly varying logarithm of P_r can be Taylor expanded about $E' = E^{(0)}$. Thus,

$$\ln \mathrm{Pr} = \ln C' + \ln \Omega'(E^{(0)}) - \left[\frac{\partial \ln \Omega'}{\partial E'}\right]_0 E_r + \cdots,$$

Note that we must expand $\ln P_r$, rather than P_r itself, because the latter function varies so rapidly with energy that the radius of convergence of its Taylor series is far too small for the series to be of any practical use. The higher order terms in previous equation can be safely neglected, because $E_r << E^{(0)}$. Now the derivative

$$\left[\frac{\partial \ln \Omega'}{\partial E'}\right]_0 \equiv \beta$$

is evaluated at the fixed energy $E' = E^{(0)}$, and is, thus, a constant independent of the energy E_r of A. In fact, we know that this derivative is just the temperature parameter $\beta = (kT)^{-1}$ characterizing the heat reservoir A'. Hence, equation becomes

$$\ln P_r = \ln C' + \ln \Omega'(E^{(0)}) - \beta E_r,$$

giving

$$P_r = C \exp(-\beta E_r),$$

where C is a constant independent of τ. The parameter C is determined by the normalization condition, which gives

$$C^{-1} = \sum_r \exp\,(-\beta E_r),$$

so that the distribution becomes

$$P_r = \frac{\exp(-\beta E_r)}{\sum_r \exp(-\beta E_r)}.$$

This is known as the Boltzmann probability distribution, and is undoubtably the most famous result in statistical physics. The Boltzmann distribution often causes confusion. People who are used to the principle of equal a priori probabilities, which says that all microstates are equally probable, are understandably surprised when they come across the Boltzmann distribution which says that high energy microstates are markedly less probable then low energy states. However, there is no need for any confusion. The principle of equal a priori probabilities applies to the whole system, whereas the Boltzmann distribution only applies to a small part of the system. The two results are perfectly consistent. If the small system is in a microstate with a comparatively high energy E_r then the rest of the system (i.e., the reservoir) has a slightly lower energy E' than usual (since the overall energy is fixed). The number of accessible microstates of the reservoir is a very strongly increasing function of its energy.

It follows that when the small system has a high energy then significantly less states than usual are accessible to the reservoir, and so the number of microstates accessible to the overall system is reduced, and, hence, the configuration is comparatively unlikely. The strong increase in the number of accessible microstates of the reservoir with increasing E' gives rise to the strong (i.e., exponential) decrease in the likelihood of a state τ of the small system with increasing E_r.

The exponential factor $\exp(-\beta\, E_r)$ is called the Boltzmann factor. The Boltzmann distribution gives the probability of finding the small system A in one particular state τ of energy E_r. The probability $P(E)$ that A has an energy in the small range between E and $E + \delta E$ is just the sum of all the probabilities of the states which lie in this range. However, since each of these states has approximately the same Boltzmann factor this sum can be written

$$P(E) = C\Omega\,(E)\exp\,(-\beta E),$$

where $\Omega(E)$is the number of microstates of A whose energies lie in the appropriate range. Suppose that system $A$is itself a large system, but still very much smaller than system A'. For a large system, we expect $\Omega(E)$ to be a very rapidly increasing function of energy, so the probability $P(E)$ is the product of a rapidly increasing function of E and another rapidly decreasing function (i.e., the Boltzmann factor).

This gives a sharp maximum of $P(E)$ at some particular value of the energy. The larger system A, the sharper this maximum becomes. Eventually, the maximum becomes so sharp that the energy of system A is almost bound to lie at the most probable energy. As usual, the most probable energy is evaluated by looking for the maximum of ln P, so

$$\frac{\partial \ln P}{\partial E} = \frac{\partial \ln \Omega}{\partial E} - \beta = 0,$$

giving

$$\frac{\partial \ln \Omega}{\partial E} = \beta.$$

Of course, this corresponds to the situation in which the temperature of A is the same as that of the reservoir. This is a result which we have seen before. Note, however, that the Boltzmann distribution is applicable no matter how small system A is, so it is a far more general result than any we have previously obtained.

SIMPLEST MICROSCOPIC SYSTEM

PARAMAGNETISM

The simplest microscopic system which we can analyze using the Boltzmann distribution is one which has only two possible states (there would clearly be little point in analyzing a system with only one possible state). Most elements, and some compounds, are paramagnetic: i.e., their constituent atoms, or molecules, possess a permanent magnetic moment due to the presence of one or more unpaired electrons. Consider a substance whose constituent particles contain only one unpaired electron. Such particles have spin 1/2, and consequently possess an intrinsic magnetic moment μ. According to quantum mechanics, the magnetic moment of a spin 1/2 particle can point either parallel or antiparallel to an external magnetic field B. Let us determine the mean magnetic moment $\bar{\mu}_B$ (in the direction of B) of the constituent particles of the substance when its absolute temperature is T. We assume, for the sake of simplicity, that each atom (or molecule) only interacts weakly with its neighbouring atoms. This enables us to focus attention on a single atom, and treat the remaining atoms as a heat bath at temperature T.

Our atom can be in one of two possible states: the (+) state in which its spin points up (i.e., parallel to B), and the (–) state in which its spin points down (i.e., antiparallel to B). In the (+)state, the atomic magnetic moment is parallel to the magnetic field, so that $\mu_\beta = \mu$. The magnetic energy of the atom is $\in_+ = - \mu B$. In the (–) state, the atomic magnetic moment is antiparallel to the magnetic field, so that $\mu_B = - \mu$. The magnetic energy of the atom is $\in_- = \mu_B$.

According to the Boltzmann distribution, the probability of finding the atom in the (+) state is

$$P_+ = C \exp(-\beta \in_+) = C \exp(\beta \, \mu \, \beta),$$

where C is a constant, and $\beta = (kT)^{-1}$. Likewise, the probability of finding the atom in the (–) state is

$$P_{-} = C\exp(-\beta \in_{-}) = C\exp(-\beta\,\mu\,B).$$

Clearly, the most probable state is the state with the lowest energy [i.e., the (+) state]. Thus, the mean magnetic moment points in the direction of the magnetic field (i.e., the atom is more likely to point parallel to the field than antiparallel).

It is clear that the critical parameter in a paramagnetic system is

$$y = \beta\mu B = \frac{\mu B}{kT}.$$

This parameter measures the ratio of the typical magnetic energy of the atom to its typical thermal energy. If the thermal energy greatly exceeds the magnetic energy then $y \ll 1$,, and the probability that the atomic moment points parallel to the magnetic field is about the same as the probability that it points antiparallel. In this situation, we expect the mean atomic moment to be small, so that $\bar{\mu}_0 \simeq 0$. On the other hand, if the magnetic energy greatly exceeds the thermal energy then $y \gg 1$, and the atomic moment is far more likely to point parallel to the magnetic field than antiparallel. In this situation, we expect $\bar{\mu}_n \simeq \mu$.

Let us calculate the mean atomic moment $\bar{\mu}_n$. The usual definition of a mean value gives

$$\bar{\mu}_B = \frac{P_+\mu + P_-(-\mu)}{P_+ + P_-} = \mu\frac{\exp(\beta\mu\beta) - \exp(-\beta\mu\beta)}{\exp(\beta\mu\beta) + \exp(-\beta\mu\beta)}.$$

This can also be written

$$\bar{\mu}_B = \mu \tan h\,\frac{\mu B}{kT},$$

where the hyperbolic tangent is defined

$$\tanh y\,\frac{\exp(y) - \exp(-y)}{\exp(y) + \exp(-y)}.$$

For small arguments $y \ll 1$,

$$\tanh y \simeq y - \frac{y^3}{3} + \cdots,$$

whereas for large arguments $y \gg 1$,

$$\tanh \simeq 1.$$

It follows that at comparatively high temperatures, $kT \gg \mu B$,

$$\bar{\mu}_B \simeq \frac{\mu^2 B}{kT},$$

whereas at comparatively low temperatures, $kT \ll \mu B$,

$$\bar{\mu}_B \simeq \mu.$$

Suppose that the substance contains N_0 atoms (or molecules) per unit volume. The magnetization is defined as the mean magnetic moment per unit volume, and is given by

$$\bar{M}_0 = N_0\, \bar{\mu}_B.$$

At high temperatures, $kT >> \mu B$, the mean magnetic moment, and, hence, the magnetization, is proportional to the applied magnetic field, so we can write

$$\bar{M}_0 \simeq \chi B,$$

where χ is a constant of proportionality known as the magnetic susceptibility. It is clear that the magnetic susceptibility of a spin 1/2 paramagnetic substance takes the form

$$\chi = \frac{N_0 \mu^2}{kT}.$$

The fact that $\chi \propto T^{-1}$ is known as Curie's law, because it was discovered experimentally by Pierre Curie at the end of the nineteenth century. At low temperatures, $kT \ll \mu B$,

$$\bar{M}_0 \to N_0 \mu,$$

so the magnetization becomes independent of the applied field. This corresponds to the maximum possible magnetization, where all atomic moments are lined up parallel to the field. The breakdown of the $\bar{M}_0 \propto B$ law at low temperatures (or high magnetic fields) is known as saturation.

The above analysis is only valid for paramagnetic substances made up of spin one-half ($J = 1/2$) atoms or molecules. However, the analysis can easily be generalized to take account of substances whose constituent particles possess higher spin (i.e., $J > 1/2$).

Compares the experimental and theoretical magnetization versus field-strength curves for three different substances made up of spin 3/2, spin 5/2, and spin 7/2 particles, showing the excellent agreement between the two sets of curves. Note that, in all cases, the magnetization is proportional to the magnetic field-strength at small field-strengths, but saturates at some constant value as the field-strength increases.

The previous analysis completely neglects any interaction between the spins of neighbouring atoms or molecules. It turns out that this is a fairly good approximation for paramagnetic substances.

However, for ferromagnetic substances, in which the spins of neighbouring atoms interact very strongly, this approximation breaks down completely. Thus, the above analysis does not apply to ferromagnetic substances.

HEAT RESERVOIR

Mean Values

Consider a system in contact with a heat reservoir. The systems in the representative ensemble are distributed over their accessible states in accordance with the Boltzmann distribution. Thus, the probability of occurrence of some state τ with energy E_r is given by

$$P_r = \frac{\exp(-\beta E_r)}{\sum_r \exp(-\beta E_r)}.$$

The mean energy is written

$$\bar{E} = \frac{\sum_r \exp(-\beta E_r) E_r}{\sum_r \exp(-\beta E_r)},$$

where the sum is taken over all states of the system, irrespective of their energy. Note that

$$\sum_r \exp(-\beta E_r) E_r = -\sum_r \frac{\partial}{\partial \beta} \exp(-\beta E_r) = -\frac{\partial Z}{\partial \beta},$$

where

$$Z - \sum_r \exp(-\beta E_r).$$

It follows that

$$\bar{E} = -\frac{1}{Z}\frac{\partial Z}{\partial \beta} = -\frac{\partial \ln Z}{\partial \beta}.$$

The quantity Z, which is defined as the sum of the Boltzmann factor over all states, irrespective of their energy, is called the partition function. We have just demonstrated that it is fairly easy to work out the mean energy of a system using its partition function. In fact, as we shall discover, it is easy to calculate virtually any piece of statistical information using the partition function.

Let us evaluate the variance of the energy. We know that

$$\overline{(\Delta E)^2} = \overline{E^2} - \overline{E}^2.$$

Now, according to the Boltzmann distribution,

$$\overline{E^2} = \frac{\sum_r \exp(-\beta E_r) E_r^2}{\sum_r \exp(-\beta E_r)}.$$

However,

$$\sum_r \exp(-\beta E_r) E_r^2 = -\frac{\partial}{\partial \beta}\left[\sum_r \exp(-E_r) E_r\right] = \left(-\frac{\partial}{\partial \beta}\right)^2 \left[\sum_r \exp(-\beta E_r)\right].$$

Hence,

$$\overline{E^2} = \frac{1}{Z}\frac{\partial^2 Z}{\partial \beta^2}.$$

We can also write

$$\overline{E^2} = \frac{\partial}{\partial \beta}\left(\frac{1}{Z}\frac{\partial Z}{\partial \beta}\right) + \frac{1}{Z^2}\left(\frac{\partial Z}{\partial \beta}\right)^2 = -\frac{\partial \overline{E}}{\partial \beta} + \overline{E}^2,$$

where use has been made of above equations. It follows from previous equation that

$$\overline{(\Delta E)^2} = -\frac{\partial \overline{E}}{\partial \beta} = \frac{\partial^2 \ln Z}{\partial \beta^2}.$$

Thus, the variance of the energy can be worked out from the partition function almost as easily as the mean energy. Since, by definition, a variance can never be negative, it follows that $\partial \overline{E}/\partial\beta \leq 0$, or, equivalently, $\partial \overline{E}/\partial T \geq 0$. Hence, the mean energy of a system governed by the Boltzmann distribution always increases with temperature.

Suppose that the system is characterized by a single external parameter x (such as its volume). The generalization to the case where there are several external parameters is obvious. Consider a quasi-static change of the external parameter from x to $x + dx$. In this process, the energy of the system in state r changes by

$$\delta E_r = -\frac{\partial E_r}{\partial x}dx.$$

The macroscopic work $đW$ done by the system due to this parameter change is

$$đW = \frac{\sum_r \exp(-E_r)(-\partial E_r/\partial x\, dx)}{\sum_r \exp(-\beta E_r)}.$$

In other words, the work done is minus the average change in internal energy of the system, where the average is calculated using the Boltzmann distribution. We can write

$$\sum_r \exp(-\beta E_r)\frac{\partial E_r}{\partial x} = -\frac{1}{\beta}\frac{\partial}{\partial x}\left[\sum_r \exp(-\beta E_r)\right] = -\frac{1}{\beta}\frac{\partial Z}{\partial x},$$

which gives

$$đW = \frac{1}{\beta Z}\frac{\partial Z}{\partial x}dx = \frac{1}{\beta}\frac{\partial \ln Z}{\partial x}dx.$$

We also have the following general expression for the work done by the system

$$đW = \overline{X}dx,$$

where

$$\overline{X} = -\overline{\frac{\partial E_r}{\partial x}}$$

is the mean generalized force conjugate to x. It follows that

$$\overline{X} = -\frac{1}{\beta}\frac{\partial \ln Z}{\partial x}.$$

Suppose that the external parameter is the volume, so $x = V$. It follows that

$$đW = \overline{p}dV = \frac{1}{\beta}\frac{\partial \ln Z}{\partial V}dV.$$

and

$$\overline{p} = \frac{1}{\beta}\frac{\partial \ln Z}{\partial V}.$$

Since the partition function is a function of β and V(the energies E_r depend on V), it is clear that the above equation relates the mean pressure $\overline{p}$ to T (via $\beta = 1/kT$) and V. In other words, the above expression is the equation of state. Hence, we can work out the pressure, and even the equation of state, using the partition function.

Partition Functions

It is clear that all important macroscopic quantities associated with a system can be expressed in terms of its partition function Z. Let us investigate how the partition function is related to thermodynamical quantities. Recall that Z is a function of both β and x (where x is the single external parameter). Hence, $Z = Z(\beta, x)$, and we can write

$$d\ln Z = \frac{\partial \ln Z}{\partial x}dx + \frac{\partial \ln Z}{\partial x}d\beta.$$

Consider a quasi-static change by which x and βchange so slowly that the system stays close to equilibrium, and, thus, remains distributed according to the Boltzmann distribution. If follows from Eqs. that

$$d\ln Z = \beta\, đW - \overline{E}\, d\beta.$$

The last term can be rewritten

$$d\ln Z = \beta đW - d(\overline{E}\,\beta) + \beta\, d\overline{E},$$

giving

$$d(\ln Z + \beta\overline{E}) = \beta(đW + d\overline{E}) \equiv \beta đQ.$$

The above equation shows that although the heat absorbed by the system $đQ$ is not an exact differential, it becomes one when multiplied by the temperature parameter β. This is essentially the second law of thermodynamics. In fact, we know that

$$dS = \frac{đQ}{T}.$$

Hence,

$$S \equiv k(\ln Z + \beta \bar{E}).$$

This expression enables us to calculate the entropy of a system from its partition function.

Suppose that we are dealing with a system $A^{(0)}$ consisting of two systems A and A'which only interact weakly with one another. Let each state of A be denoted by an index $\tau$and have a corresponding energy E_r. Likewise, let each state of A' be denoted by an index s and have a corresponding energy $E's$. A state of the combined system $A^{(0)}$ is then denoted by two indices τ and s. Since A and A' only interact weakly their energies are additive, and the energy of state τs is

$$E_{\tau s}^{(0)} = E_r + E_s'.$$

By definition, the partition function of $A^{(0)}$ takes the form

$$Z^{(0)} = \sum_{\tau,s} \exp[-\beta\, E_{\tau s}^{(0)}]$$

$$= \sum_{\tau,s} \exp(-\beta\, [E_r + E_s'])$$

$$= \left[\sum_{\tau} \exp(-\beta\, E_r)\right]\left[\sum_{s} \exp(-\beta\, E_s')\right].$$

Hence,

$$Z^{(0)} = ZZ',$$

giving

$$\ln Z^{(0)} = \ln Z + \ln Z',$$

where Z and Z' are the partition functions of A and A', respectively. It follows from equation. that the mean energies of $A^{(0)}$, A, and A' are related by

$$\bar{E}^{(0)} = \bar{E} + \bar{E}'.$$

It also follows from equation that the respective entropies of these systems are related via

$$S^{(0)} = S + S'.$$

Hence, the partition function tells us that the extensive thermodynamic functions of two weakly interacting systems are simply additive.

It is clear that we can perform statistical thermodynamical calculations using the partition function Zinstead of the more direct approach in which we use the density of states Ω. The former approach is advantageous because the partition function is an unrestricted sum of Boltzmann factors over all accessible states, irrespective of their energy, whereas the density of states is a restricted sum over all states whose energies lie in some narrow range.

In general, it is far easier to perform an unrestricted sum than a restricted sum. Thus, it is generally easier to derive statistical thermodynamical results using Zrather than Ω, although Ω has a far more direct physical significance than Z.

THERMODYNAMIC RELATIONS

IDEAL MONATOMIC GASES

Let us now practice calculating thermodynamic relations using the partition function by considering an example with which we are already quite familiar: i.e., an ideal monatomic gas. Consider a gas consisting of Nidentical monatomic molecules of mass m enclosed in a container of volume V. Let us denote the position and momentum vectors of the ith molecule by r_i and p_i, respectively. Since the gas is ideal, there are no interatomic forces, and the total energy is simply the sum of the individual kinetic energies of the molecules:

$$E = \sum_{i=1}^{N} \frac{{p_i}^2}{2m},$$

where $p_i^2 = p_{i.}\, p_i$.

Let us treat the problem classically. In this approach, we divide up phase-space into cells of equal volume h_0^f. Here, f is the number of degrees of freedom, and h_0 is a small constant with dimensions of angular momentum which parameterizes the precision to which the positions and momenta of molecules are determined. Each cell in phase-space corresponds to a different state. The partition function is the sum of the Boltzmann factor exp $(-\beta\, E_r)$ over all possible states, where E_r is the energy of state τ. Classically, we can approximate the summation over cells in phase-space as an integration over all phase-space. Thus,

$$Z = \int \cdots \int \exp(-\beta\, E) \frac{d^3 r_1 \cdots d^3 r_N\; d^3 p_1 \cdots d^3{}_{PN}}{{h_0}^{3N}},$$

where $3N$ is the number of degrees of freedom of a monatomic gas containing N molecules. Making use of equationthe above expression reduces to

$$Z = \frac{V^N}{{h_0}^{3N}} \int \cdots \int \exp[-\beta/2m) {p_1}^2]\, d^3 p_1 \cdots \exp[-(\beta/2m) p_N^2] d^3{}_{PN}.$$

Note that the integral over the coordinates of a given molecule simply yields the volume of the container, V, since the energy E is independent of the locations of the molecules in an ideal gas. There are N such integrals, so we obtain the factor V^N in the above expression. Note, also, that each of the integrals over the molecular momenta in equation are identical: they differ only by irrelevant dummy variables of integration. It follows that the partition function Z of the gas is made up of the product of N identical factors: i.e.,

$$Z = \zeta^N,$$

where

$$\zeta = \frac{V}{{h_0}^3} \int \exp[-(\beta/2m)p^2] d^3p$$

is the partition function for a single molecule. Of course, this result is obvious, since we have already shown that the partition function for a system made up of a number of weakly interacting subsystems is just the product of the partition functions of the subsystems.

The integral in equation is easily evaluated:

$$\int \exp[-(\beta/2m)p^2] d^3p = \int_{-\infty}^{\infty} \exp[-(\beta/2m)\, {p_x}^2] dp_x$$

$$\int_{-\infty}^{\infty} \exp[-(\beta/2m)\, {p_y}^2] dp_y \times \int_{-\infty}^{\infty} \exp[-(\beta/2m)\, {p_z}^2] dp_z$$

$$= \left(\sqrt{\frac{2\pi m}{\beta}}\right)^3,$$

where use has been made of equation. Thus,

$$\zeta = V\left(\frac{2\pi m}{{h_0}^2 \beta}\right)^{3/2},$$

and

$$\ln Z = N \ln \zeta = N\left[\ln V - \frac{3}{2}\ln\beta + \frac{3}{2}\ln\left(\frac{2\pi m}{{h_0}^2}\right)\right].$$

The expression for the mean pressure yields

$$\bar{p} = \frac{1}{\beta}\frac{\partial \ln Z}{\partial V} = \frac{1}{\beta}\frac{N}{V},$$

which reduces to the ideal gas equation of state

$$pV = NkT = \upsilon RT,$$

where use has been made of $N = \upsilon N_A$ and $R = N_A k$. According to equation the mean energy of the gas is given by

$$\bar{E} = -\frac{\partial \ln Z}{\partial \beta} = \frac{3}{2}\frac{N}{\beta} = \upsilon\frac{3}{2}RT.$$

Note that the internal energy is a function of temperature alone, with no dependence on volume. The molar heat capacity at constant volume of the gas is given by

$$c_v = \frac{1}{v}\left(\frac{\partial \bar{E}}{\partial T}\right)_V = \frac{3}{2}R,$$

so the mean energy can be written

$$\bar{E} = v\, c_v T.$$

We have seen all of the above results before. Let us now use the partition function to calculate a new result. The entropy of the gas can be calculated quite simply from the expression

$$S = k(\ln Z + \beta \bar{E}).$$

Thus,

$$S = vR\left[\ln V - \frac{3}{2}\ln\beta + \frac{3}{2}\ln\left(\frac{2\pi m}{h_0^{\,2}}\right) + \frac{3}{2}\right],$$

or

$$S = vR\left[\ln V + \frac{3}{2}\ln T + \sigma\right],$$

where

$$\sigma = \frac{3}{2}\ln\left(\frac{2\pi m k}{h_0^{\,2}}\right) + \frac{3}{2}.$$

The above expression for the entropy of an ideal gas is certainly new. Unfortunately, it is also quite obviously incorrect.

GIBB'S PARADOX

What has gone wrong? First of all, let us be clear why equation is incorrect. We can see that $S \to -\infty$ as $T \to 0$, which contradicts the third law of thermodynamics. However, this is not a problem. Equation was derived using classical physics, which breaks down at low temperatures. Thus, we would not expect this equation to give a sensible answer close to the absolute zero of temperature.

Equation is wrong because it implies that the entropy does not behave properly as an extensive quantity. Thermodynamic quantities can be divided into two groups, extensive and intensive. Extensive quantities increase by a factor α when the size of the system under consideration is increased by the same factor. Intensive quantities stay the same. Energy and volume are typical extensive quantities. Pressure and temperature are typical intensive quantities. Entropy is very definitely an extensive quantity. We have shown that the entropies of two weakly interacting systems are additive. Thus, if we double the size of a system we expect the entropy to double as well. Suppose that we have a system of volume V containing v moles of ideal gas at temperature T. Doubling the size of the system is like joining two identical systems together to form a new system of volume $2V$ containing $2v$ moles of gas at temperature T. Let

$$S = vR\left[\ln V + \frac{3}{2}\ln T + \sigma\right]$$

denote the entropy of the original system, and let

$$S' = 2vR\left[\ln 2V + \frac{3}{2}\ln T + \sigma\right]$$

denote the entropy of the double-sized system. Clearly, if entropy is an extensive quantity (which it is!) then we should have

$$S' = 2S.$$

But, in fact, we find that

$$S' - 2S = 2vR\ln 2.$$

So, the entropy of the double-sized system is more than double the entropy of the original system. Where does this extra entropy come from? Well, let us consider a little more carefully how we might go about doubling the size of our system. Suppose that we put another identical system adjacent to it, and separate the two systems by a partition. Let us now suddenly remove the partition.

If entropy is a properly extensive quantity then the entropy of the overall system should be the same before and after the partition is removed. It is certainly the case that the energy (another extensive quantity) of the overall system stays the same.

However, according to above equation the overall entropy of the system increases by $2vR$ ln 2 after the partition is removed. Suppose, now, that the second system is identical to the first system in all respects except that its molecules are in some way slightly different to the molecules in the first system, so that the two sets of molecules are distinguishable. In this case, we would certainly expect an overall increase in entropy when the partition is removed. Before the partition is removed, it separates type 1 molecules from type 2 molecules. After the partition is removed, molecules of both types become jumbled together.

This is clearly an irreversible process. We cannot imagine the molecules spontaneously sorting themselves out again. The increase in entropy associated with this jumbling is called entropy of mixing, and is easily calculated. We know that the number of accessible states of an ideal gas varies with volume like $\Omega \propto V^N$.

The volume accessible to type 1 molecules clearly doubles after the partition is removed, as does the volume accessible to type 2 molecules.

Using the fundamental formula $S = k \ln \Omega$, the increase in entropy due to mixing is given by

$$S = 2k \ln \frac{\Omega_f}{\Omega_i} = 2Nk \ln \frac{V_f}{V_i} = 2v\, R \ln 2.$$

It is clear that the additional entropy $2vR \ln 2$, which appears when we double the size of an ideal gas system by joining together two identical systems, is entropy of mixing of the molecules contained in the original systems. But, if the molecules in these two systems are indistinguishable, why should there be any entropy of mixing? Well, clearly, there is no entropy of mixing in this case.

At this point, we can begin to understand what has gone wrong in our calculation. We have calculated the partition function assuming that all of the molecules in our system have the same mass and temperature, but we have never explicitly taken into account the fact that we consider the molecules to be indistinguishable.

In other words, we have been treating the molecules in our ideal gas as if each carried a little license plate, or a social security number, so that we could always tell one from another. In quantum mechanics, which is what we really should be using to study microscopic phenomena, the essential indistinguishability of atoms and molecules is hard-wired into the theory at a very low level. Our problem is that we have been taking the classical approach a little too seriously. It is plainly silly to pretend that we can distinguish molecules in a statistical problem, where we do not closely follow the motions of individual particles. A paradox arises if we try to treat molecules as if they were distinguishable. This is called Gibb's paradox, after the American physicist Josiah Gibbs who first discussed it. The resolution of Gibb's paradox is quite simple: treat all molecules of the same species as if they were indistinguishable. In our previous calculation of the ideal gas partition function, we inadvertently treated each of the N molecules in the gas as distinguishable.

Because of this, we overcounted the number of states of the system. Since the $N!$ possible permutations of the molecules amongst themselves do not lead to physically different situations, and, therefore, cannot be counted as separate states, the number of actual states of the system is a factor $N!$ less than what we initially thought. We can easily correct our partition function by simply dividing by this factor, so that

$$Z = \frac{\zeta^N}{N!}.$$

This gives

$$\ln Z = N \ln \zeta - N!,$$

or

$$\ln Z = N \ln \zeta - N \ln N + N,$$

using Stirling's approximation. Note that our new version of ln Z differs from our previous version by an additive term involving the number of particles in the system. This explains why our calculations of the mean pressure and mean energy, which depend on partial derivatives of ln Z with respect to the volume and the temperature parameter β, respectively, came out all right. However, our expression for the entropy S is modified by this additive term. The new expression is

$$S = \upsilon R\left[\ln V - \frac{3}{2}\ln\beta + \frac{3}{2}\ln\left(\frac{2\pi\, mk}{h_0^{\,2}}\right) + \frac{3}{2}\right] + k(-N\ln N + N).$$

This gives

$$S = \upsilon R\left[\ln\frac{V}{N} + \frac{3}{2}\ln T + \sigma_0\right]$$

where

$$\sigma_0 = \frac{3}{2}\ln\left(\frac{2\pi\, mk}{h_0^{\,2}}\right) + \frac{5}{2}.$$

It is clear that the entropy behaves properly as an extensive quantity in the above expression: i.e., it is multiplied by a factor α when υ, V, and N are multiplied by the same factor.

THE MAXWELL DISTRIBUTION

Consider a molecule of mass m in a gas which is sufficiently dilute for the intermolecular forces to be negligible (i.e., an ideal gas). The energy of the molecule is written

$$\in = \frac{p^2}{2m} + \in^{\text{int}},$$

where p is its momentum vector, and $\in^{\text{int}}$ is its internal (i.e., non-translational) energy. The latter energy is due to molecular rotation, vibration, etc.

Translational degrees of freedom can be treated classically to an excellent approximation, whereas internal degrees of freedom usually require a quantum mechanical approach.

Classically, the probability of finding the molecule in a given internal state with a position vector in the range r to $r + dr$, and a momentum vector in the range P to $P + dP$, is proportional to the number of cells (of "volume" h_0) contained in the corresponding region of phase-space, weighted by the Boltzmann factor.

In fact, since classical phase-space is divided up into uniform cells, the number of cells is just proportional to the "volume" of the region under consideration. This "volume" is written $d^3r d^3p$. Thus, the probability of finding the molecule in a given internal state s is

$$P_s(r,p)d^3r\, d^3p \propto \exp(-\beta p^2/2m)\exp(\beta\in_s^{\text{int}})d^3r\, d^3p,$$

where Ps is a probability density defined in the usual manner. The probability $P(r, p)\, d^3r\, d^3p$ of finding the molecule in any internal state with position and momentum vectors in the specified range is obtained by summing the above expression over all possible internal states.

The sum over $\exp(-\beta \in_s^{int})$ just contributes a constant of proportionality (since the internal states do not depend on r or p), so

$$P(r,p)d^3r\, d^3p \propto \exp(-\beta p^2/2m)\, d^3r\, d^3p.$$

Of course, we can multiply this probability by the total number of molecules N in order to obtain the mean number of molecules with position and momentum vectors in the specified range.

Suppose that we now want to determine:$f(r, v)\, d^3r\, d^3v$ i.e., the mean number of molecules with positions between r and $r + dr$, and velocities in the range v and $v + dv$. Since $v = p/m$, it is easily seen that

$$f(r,v)d^3r\, d^3v = C\exp(-\beta mv^2/2)\, d^3r\, d^3v,$$

where C is a constant of proportionality. This constant can be determined by the condition

$$\int_{(r)}\int_{(v)} f(r,v)d^3r\, d^3v = N:$$

i.e., the sum over molecules with all possible positions and velocities gives the total number of molecules, N. The integral over the molecular position coordinates just gives the volume V of the gas, since the Boltzmann factor is independent of position. The integration over the velocity coordinates can be reduced to the product of three identical integrals (one for v_x, one for v_y, and one for v_z), so we have

$$CV\left[\int_{-\infty}^{\infty}\exp(-\beta mv_z^2/2)dv_z\right]^3 = N.$$

Now,

$$\int_{-\infty}^{\infty}\exp(-\beta mv_z^2/2)dv_z = \sqrt{\frac{2}{\beta m}}\int_{-\infty}^{\infty}\exp(-y^2)dy = \sqrt{\frac{2\pi}{\beta m}},$$

so $C = (N/V)\,(\beta m/2\pi)^{3/2}$. Thus, the properly normalized distribution function for molecular velocities is written

$$f(v)d^3r\, d^3v = n\left(\frac{m}{2\pi kT}\right)^{3/2}\exp(-mv^2/2kT)d^3r\, d^3v.$$

Here, $n = N/V$ is the number density of the molecules. We have omitted the variable r in the argument of f, since f clearly does not depend on position. In other words, the distribution of molecular velocities is uniform in space. This is hardly surprising, since there is nothing to distinguish one region of space from another in our calculation.

The above distribution is called the Maxwell velocity distribution, because it was discovered by James Clark Maxwell in the middle of the nineteenth century. The average number of molecules per unit volume with velocities in the range v to $v + dv$ is obviously $f(v)\,d^3v$.

Let us consider the distribution of a given component of velocity: the z-component, say. Suppose that $g(v_z)\,dv_z$ is the average number of molecules per unit volume with the z-component of velocity in the range v_z to $v_z + dv_z$, irrespective of the values of their other velocity components.

It is fairly obvious that this distribution is obtained from the Maxwell distribution by summing (integrating actually) over all possible values of v_x and v_y, with v_z in the specified range. Thus,

$$g(vz)dv_z = \int_{(v_z)}\int_{(v_y)} f(v)d^3v.$$

This gives

$$g(v_z)dv_z = n\left(\frac{m}{2\pi kT}\right)$$

$$\int_{(v_z)}\int_{(v_y)} \exp\left[-(m/2kT)\left(v_z^2 + v_v^2 + v_z^2\right)\right]dv_x dv_y dv_z$$

$$= n\left(\frac{m}{2\pi\, kT}\right)^{3/2} \exp\left(-mv_z^2/2kT\right)\left[\int_{-\infty}^{\infty} \exp\left(-mv_x^2/2k\,T\right)\right]^2$$

$$= n\left(\frac{m}{2\pi\, kT}\right)^{3/2} \exp\left(-mv_z^2/2\,kT\right)\left(\sqrt{\frac{2\pi\, kT}{m}}\right)^2,$$

or

$$g(v_z)\,dv_z = n\left(\frac{m}{2\pi\, kT}\right)^{1/2} \exp\left(-mv_z^2/2\,kT\right)dv_z.$$

Of course, this expression is properly normalized, so that

$$\int_{-\infty}^{\infty} g(v_z)dv_z = n.$$

It is clear that each component (since there is nothing special about the z-component) of the velocity is distributed with a Gaussian probability distribution, centred on a mean value

$$\overline{v_z} = 0,$$

with variance

$$\overline{v_z^2} = \frac{kT}{m}.$$

Equation implies that each molecule is just as likely to be moving in the plus z-direction as in the minus z-direction. Equation can be rearranged to give

$$\overline{\frac{1}{2}mv_z^2} = \frac{1}{2}kT,$$

in accordance with the equipartition theorem.

Note that equation can be rewritten

$$\frac{f(v)d^3v}{n} = \left[\frac{g(v_x)dv_x}{n}\right]\left[\frac{g(v_y)dv_y}{n}\right]\left[\frac{g(v_z)dv_x}{n}\right],$$

where $g(v_x)$ and $g(v_y)$ are defined in an analogous way to $g(v_z)$. Thus, the probability that the velocity lies in the range v to $v + dv$ is just equal to the product of the probabilities that the velocity components lie in their respective ranges. In other words, the individual velocity components act like statistically independent variables.

Suppose that we now want to calculate $F(v)\,dv$: i.e., the average number of molecules per unit volume with a speed $v = |v|$ in the range v to $v + dv$. It is obvious that we can obtain this quantity by adding up all molecules with speeds in this range, irrespective of the direction of their velocities. Thus,

$$F(v)dv = \int f(v)d^3v,$$

where the integral extends over all velocities satisfying

$$v < |v| < v + dv.$$

This inequality is satisfied by a spherical shell of radius v and thickness dv in velocity space. Since $f(v)$ only depends on $|v|$, so $f(v) \equiv f(v)$, the above integral is just $f(v)$ multiplied by the volume of the spherical shell in velocity space. So,

$$F(v)dv = 4\pi f(v)v^2 dv,$$

which gives

$$F(v)dv = 4\pi f(v)\left(\frac{m}{2\pi\, kT}\right)v^2 \exp(-mv^2 / 2k\,T)\,dv.$$

This is the famous Maxwell distribution of molecular speeds. Of course, it is properly normalized, so that

$$\int_0^\infty F(v)dv = n.$$

Note that the Maxwell distribution exhibits a maximum at some non-zero value of v. The reason for this is quite simple. As v increases, the Boltzmann factor decreases, but the volume of phase-space available to the molecule (which is proportional to v^2) increases: the net result is a distribution with a non-zero maximum.

The mean molecular speed is given by

$$\bar{v} = \frac{1}{n}\int_0^\infty F(v)v\,dv.$$

Thus, we obtain

$$\bar{v} = 4\pi\left(\frac{m}{2\pi\,kT}\right)^{3/2}\int_0^\infty v^3\exp(-mv^2/2kT)\,dv,$$

or

$$\bar{v} = 4\pi\left(\frac{m}{2\pi\,kT}\right)^{3/2}\left(\frac{2k\,T}{m}\right)^2\int_0^\infty y^3\exp(-y^2)\,dy$$

Now

$$\int_0^\infty y^3\exp(-y^2)\,dy = \frac{1}{2},$$

so

$$\bar{v} = \sqrt{\frac{8}{\pi}\frac{kT}{m}}.$$

A similar calculation gives

$$v_{\text{rms}} = \sqrt{\overline{v^2}} = \sqrt{\frac{3\,kT}{m}}.$$

However, this result can also be obtained from the equipartition theorem. Since

$$\overline{\frac{1}{2}mv^2} = \overline{\frac{1}{2}m\left(v_z^2 + v_y^2 + v_z^2\right)} = 3\left(\frac{1}{2}kT\right),$$

then equation follows immediately. It is easily demonstrated that the most probable molecular speed (i.e., the maximum of the Maxwell distribution function) is

$$\bar{v} = \sqrt{\frac{2kT}{m}}.$$

The speed of sound in an ideal gas is given by

$$c_s = \sqrt{\frac{\gamma\,kT}{m}},$$

where γ is the ratio of specific heats. This can also be written since $p = nkT$ and $\rho = nm$. It is clear that the various average speeds which we have just calculated are all of order the sound speed (i.e., a few hundred meters per second at room temperature). In ordinary air ($\gamma = 1.4$) the sound speed is about 84% of the most probable molecular speed, and about 74% of the mean molecular speed. Since sound waves ultimately propagate via molecular motion, it makes sense that they travel at slightly less than the most probable and mean molecular speeds.

The Maxwell velocity distribution as a function of molecular speed in units of the most probable speed. Also shown are the mean speed and the root mean square speed.

It is difficult to directly verify the Maxwell velocity distribution. However, this distribution can be verified indirectly by measuring the velocity distribution of atoms exiting from a small hole in an oven.

The velocity distribution of the escaping atoms is closely related to, but slightly different from, the velocity distribution inside the oven, since high velocity atoms escape more readily than low velocity atoms. In fact, the predicted velocity distribution of the escaping atoms varies like $v^3 \exp(-mv^2/2kT)$, in contrast to the $v^3 \exp(-mv^2/2kT)$ variation of the velocity distribution inside the oven. The measured and theoretically predicted velocity distributions of potassium atoms escaping from an oven at 157 °C. There is clearly very good agreement between the two.

3

Thermodynamic Systems

An important concept in thermodynamics is the "system". Everything in the universe except the system is known as surroundings. A system is the region of the universe under study. A system is separated from the remainder of the universe by a boundary which may be imaginary or not, but which by convention delimits a finite volume. The possible exchanges of work, heat, or matter between the system and the surroundings take place across this boundary. Boundaries are of four types: fixed, moveable, real, and imaginary.

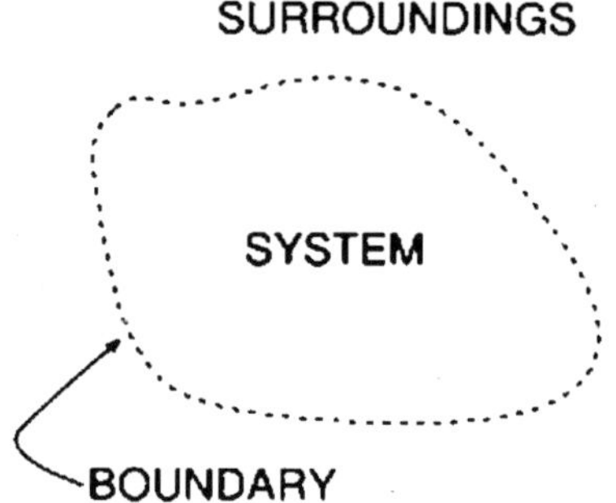

Fig. Thermodynamic Systems

Basically, the "boundary" is simply an imaginary dotted line drawn around a volume of something when there is going to be a change in the internal energy of that something. Anything that passes across the boundary that effects a change in the internal energy of the something needs to be accounted for in the energy balance equation. That something can be the volumetric region surrounding a single atom resonating energy, such as Max Planck defined in 1900; it can be a body of steam or air in a steam engine, such as Sadi Carnot defined in 1824; it can be the body of a tropical cyclone, such as Kerry Emanuel theorized in 1986 in the field of atmospheric thermodynamics; it could also be just one nuclide (i.e. a system of quarks) as some are theorizing presently in quantum thermodynamics.

For an engine, a fixed boundary means the piston is locked at its position; as such, a constant volume process occurs. In that same engine, a moveable boundary allows the piston to move in and out. For closed systems, boundaries are real while for open system boundaries are often imaginary. There are five dominant classes of systems:

- *Isolated Systems*–matter and energy may not cross the boundary
- *Adiabatic Systems*–heat must not cross the boundary
- *Diathermic Systems*–heat may cross boundary
- *Closed Systems*–matter may not cross the boundary
- *Open Systems*–heat, work, and matter may cross the boundary

As time passes in an isolated system, internal differences in the system tend to even out and pressures and temperatures tend to equalize, as do density differences. A system in which all equalizing processes have gone practically to completion, is considered to be in a state of thermodynamic equilibrium.

In thermodynamic equilibrium, a system's properties are, by definition, unchanging in time. Systems in equilibrium are much simpler and easier to understand than systems which are not in equilibrium. Often, when analysing a thermodynamic process, it can be assumed that each intermediate state in the process is at equilibrium. This will also considerably simplify the situation. Thermodynamic processes which develop so slowly as to allow each intermediate step to be an equilibrium state are said to be reversible processes.

CLOSED AND OPEN SYSTEM OF THERMODYNAMIC SYSTEM

In thermodynamics, a thermodynamic system, originally called a working substance, is defined as that part of the universe that is under consideration.

A hypothetical boundary separates the system from the rest of the universe, which is referred to as the environment, surroundings, or reservoir. A useful classification of thermodynamic systems is based on the nature of the boundary and the quantities flowing through it, such as matter, energy, work, heat, and entropy. A system can be anything, for example a piston, a solution in a test tube, a living organism, an electrical circuit, a planet, etc.

Thermodynamics is conducted under a *system-centred view* of the universe. All quantities, such as pressure or mechanical work, in an equation refer to the system unless labeled otherwise. Thermodynamics is basically concerned with the flow and balance of energy and matter in a thermodynamic system. Three types of thermodynamic systems are distinguished depending on the kinds of interaction and energy exchange taking place between the system and its surrounding environment:

- Isolated systems are completely isolated in every way from their environment. They do not exchange heat, work or matter with their environment. An example of an isolated system would be an insulated rigid container, such as an insulated gas cylinder.
- Closed systems are able to exchange energy (heat and work) but not matter with their environment. A greenhouse is an example of

a closed system exchanging heat but not work with its environment. Whether a system exchanges heat, work or both is usually thought of as a property of its boundary.

- Open systems: exchanging energy (heat and work) and matter with their environment. A boundary allowing matter exchange is called *permeable.* The ocean would be an example of an open system.

In reality, a system can never be absolutely isolated from its environment, because there is always at least some slight coupling, even if only via minimal gravitational attraction. In analyzing a system in steady-state, the energy into the system is equal to the energy leaving the system.

As an example, consider the system of hot liquid water and solid table salt in a sealed, insulated test tube held in a vacuum (the surroundings). The test tube constantly loses heat (in the form of black-body radiation), but the heat loss progresses very slowly. If there is another process going on in the test tube, for example the dissolution of the salt crystals, it will probably occur so quickly that any heat lost to the test tube during that time can be neglected. (Thermodynamics does not measure time, but it does sometimes accept limitations on the timeframe of a process.)

HISTORY

The first to develop the concept of a "thermodynamic system" was the French physicist Sadi Carnot whose 1824 *Reflections on the Motive Power of Fire* studied what he called the "working substance" (system), i.e. typically a body of water vapour, in steam engines, in regards to the system's ability to do work when heat is applied to it. The working substance could be put in contact with either a heat reservoir (a boiler), a cold reservoir (a stream of cold water), or a piston (to which the working body could do work by pushing on it). In 1850, the German physicist Rudolf Clausius generalized this picture to include the concept of the surroundings, and began referring to the system as a "working body." In his 1850 manuscript *On the Motive Power of Fire,* Clausius wrote:

"With every change of volume (to the working body) a certain amount work must be done by the gas or upon it, since by its expansion it overcomes an external pressure, and since its compression can be brought about only by an exertion of external pressure.

To this excess of work done by the gas or upon it there must correspond, by our principle, a proportional excess of heat consumed or produced, and the gas cannot give up to the "surrounding medium" the same amount of heat as it receives."

The article Carnot heat engine shows the original piston-and-cylinder diagram used by Carnot in discussing his ideal engine; below, we see the Carnot engine as is typically modeled in current use: Carnot engine diagram (modern) - where heat flows from a high temperature T_H furnace through the fluid of the "working body" (working substance) and into the cold sink T_C, thus forcing the working substance to do mechanical work W on the surroundings, via cycles of contractions and expansions.

In the diagram shown, the "working body" (system), a term introduced by Clausius in 1850, can be any fluid or vapour body through which heat Q can be introduced or transmitted through to produce work. In 1824, Sadi Carnot, in his famous paper *Reflections on the Motive Power of Fire*, had postulated that the fluid body could be any substance capable of expansion, such as vapour of water, vapour of alcohol, vapour of mercury, a permanent gas, or air, etc. Although, in these early years, engines came in a number of configurations, typically Q_H was supplied by a boiler, wherein water was boiled over a furnace; Q_C was typically a stream of cold flowing water in the form of a condenser located on a separate part of the engine. The output work W here is the movement of the piston as it is used to turn a crank-arm, which was then typically used to turn a pulley so to lift water out of flooded salt mines. Carnot defined work as "weight lifted through a height."

SYSTEMS IN EQUILIBRIUM

In isolated systems it is consistently observed that as time goes on internal rearrangements diminish and stable conditions are approached. Pressures and temperatures tend to equalize, and matter arranges itself into one or a few relatively homogeneous phases. A system in which all processes of change have gone practically to completion is considered to be in a state of thermodynamic equilibrium. The thermodynamic properties of a system in equilibrium are unchanging in time. Equilibrium system states are much easier to describe in a deterministic manner than non-equilibrium states.

In thermodynamic processes, large departures from equilibrium during intermediate steps are associated with increases in entropy and increases in the production of heat rather than useful work. It can be shown that for a process to be reversible, each step in the process must be reversible. For a step in a process to be reversible, the system must be in equilibrium throughout the step. That ideal cannot be accomplished in practice because no step can be taken without perturbing the system from equilibrium, but the ideal can be approached by making changes slowly.

OPEN SYSTEMS

In open systems, matter may flow in and out of the system boundaries. The first law of thermodynamics for open systems states: the increase in the internal energy of a system is equal to the amount of energy added to the system by matter flowing in and by heating, minus the amount lost by matter flowing out and in the form of work done by the system. The first law for open systems is given by: During steady, continuous operation, an energy balance applied to an open system equates shaft work performed by the system to heat added plus net enthalpy added. where U_{in} is the average internal energy entering the system and U_{out} is the average internal energy leaving the system. The region of space enclosed by open system boundaries is usually called a control volume, and it may or may not correspond to physical walls. If we choose the shape of the control volume such that all flow in or out occurs perpendicular to its surface, then the flow of matter into the system performs work as if it were a piston of fluid pushing mass into the system, and the system performs work on the flow of matter out as if it were driving a piston of fluid. There are then two types of work performed: *flow work* described above which is performed on the fluid (this is also often called *PV work*) and *shaft work* which may be performed on some mechanical device. These two types of work are expressed in the equation:

Substitution into the equation above for the control volume *cv* yields: The definition of enthalpy, *H*, permits us to use this thermodynamic potential to account for both internal energy and PV work in fluids for open systems:

THERMODYNAMIC POTENTIALS

Most real thermodynamic systems are open systems that exchange heat and work with their environment, rather than the closed systems described thus far. For example, living systems are clearly able to achieve a local reduction in their entropy as they grow and develop; they create structures of greater internal energy (i.e., they lower entropy) out of the nutrients they absorb. This does not represent a violation of the second law of thermodynamics, because a living organism does not constitute a closed system.

In order to simplify the application of the laws of thermodynamics to open systems, parametres with the dimensions of energy, known as thermodynamic potentials, are introduced to describe the system. The resulting formulas are expressed in terms of the Helmholtz free energy *F* and the Gibbs free energy *G*, named after the 19th-century German physiologist and physicist Hermann von Helmholtz and the contemporaneous American physicist Josiah Willard Gibbs. The key conceptual step is to separate a system from its heat reservoir. A system is thought of as being held at a constant temperature *T* by a heat reservoir (i.e., the environment), but the heat reservoir is no longer considered to be part of the system.

EFFECTS OF FLOW AND NON-FLOW PROCESSES

Anionic surfactants, good foaming agents, can be used in CO_2 foam flooding to improve high-pressure, high-density CO_2 reservoir sweep efficiency. In this study, kinetics and equilibrium adsorption were investigated by examining adsorption behaviour in a system of solid phase sandstone or limestone and of an aqueous phase of surfactant in 2% brine. Effects on surfactant adsorption density for different solid to liquid ratios as well as surfactant concentration, rock type and state, and flow conditions are presented. Three systems were used: batch tests on crushed rock, circulation tests through core samples, and non-flow, diffusion brine-saturated core tests.

The density of an anionic surfactant adsorption on rock is best described as a function of surfactant available in the system, rather than by surfactant concentration used by previous investigators. Experiments with solid rock were carried out to determine surfactant capacity of the porous media through flow and non-flow rock samples and crushed rock samples. Adsorption was similar for crushed sandstone and flow tests, while significantly higher in cubed, non-flow rock systems. For limestone the crushed rock and non-flow systems were similar while the flow system was significantly lower. The time to reach equilibrium required less than one hour (generally minutes) for the crushed rock, hours to days for the flow-through tests, and weeks to over a month for the non-flow rock systems. The rate of adsorption dependent on availability (delivery) that is generally much slower than the adsorption kinetics.

INTRODUCTION

Surfactants are widely used in a large number of applications because of their remarkable ability to influence the properties of surfaces and interfaces. The applications of surfactants in petroleum industry are diverse and significant. Some of these areas include: in situ, wellbores, surface facilities, and environmental, health, and safety applications.[1] In each case appropriate knowledge and practices determine both the economic and technical successes of the industrial process concerned.

In-situ CO_2 flooding processes frequently experience poor sweep efficiency despite the favourable characteristics of CO_2 in recovery of oil. To mitigate this problem, significant research has focused on the use of foam in CO_2 flooding processes. It has been established that foam can improve sweep efficiency in CO_2 flooding processes by reducing the mobility of CO_2 and by diverting CO_2 flow to previously bypassed zones.[2-8]

Foam application involves injecting a surfactant along with water and gas into the reservoir. Normally, the surfactant is dissolved in the aqueous phase though surfactant dissolved in CO_2 are also being considered.[9] To aid in the selection of a suitable surfactant for the reservoir, laboratory data is collected to characterize the surfactant performance.

The economics of foam flooding depend significantly on the quantity of surfactant required to generate and propagate foam. Surfactant loss through partitioning into the crude oil phase and through adsorption onto the rock surfaces often consumes more than 90% of the surfactant in the system.[5] Surfactant loss through partitioning into the crude oil can be responsible for surfactant losses of as much as 30%. However, for the very hydrophilic surfactants chosen for many foam flooding applications, the partitioning onto crude oil is near zero.[1] More serious losses are demonstrated from the results of a number of studies of the adsorption properties of surfactants suitable for foam flooding. These have shown that effective foam forming surfactant may exhibit adsorption levels from near zero to up to quite high levels on the order.

For example one test determined adsorption rates of 2.5 mg/g (mg of adsorbed surfactant for each gram of rock in the system). A mile-square, 10 feet thick reservoir saturated with surfactant would require 55,000 tons (50,000 tones) of surfactant at 2.5 mg of surfactant per g of rock. Adsorption does not depend on the nature of the surfactant alone, but also (though not inclusively) on temperature, brine salinity and hardness, rock type, wettability, and the presence of the residual oil phase. These factors will lead to vastly different distances of foam propagation into a reservoir, so selection of foam-forming surfactant formulation with acceptable adsorption levels at reservoir conditions is critical.

Soil respiration estimates obtained from non-flow-through steady-state chambers (also called static, absorption, or alkali trap chambers) are considered by many investigators to be unreliable. We studied the accuracy, functioning, and design requirements of this chamber type using a gas diffusion model validated for this purpose by demonstrating that it matched the empirical relation between alkali-measured flux and headspace CO_2 concentration.

Simulated measurement error depended on:

(i) Magnitude of the soil respiration rate, which spawned positive or negative error depending on the algebraic sign of the change in headspace CO_2,

(ii) Absorption efficiency of the alkali trap, which was determined by headspace air mixing rates, the thickness of atmospheric interfacial layers, and especially the ratio of exposed alkali surface area to emitting soil surface area,

(iii) The effective diffusivity and storage coefficient of CO_2 in underlying soil, which depended on the soil's air-filled porosity (AFP) and pH, respectively,

(iv) The rate of CO_2 leakage between the chamber system and its surroundings.

The results also indicated that although no single chamber design is universally applicable, striving for the ideal design in every situation is not required; for example, measurement error associated with the design used in our simulations was usually only 5% despite that headspace concentration rose more than 70% within 2 h. Larger errors occurred for chamber designs less well matched to the soil respiration rate they were intended to measure, but if such serious design deficiencies are avoided, the method offers a simple inexpensive means for obtaining multiple reliable time-integrated estimates of soil respiration, even at remote locations.

Abbreviations

AFP, air-filled porosity (m^3 m^{-3} soil):

- *D*, binary molecular diffusion coefficient of CO_2 in air
- F_a, estimate of CO_2 flux density at the soil-atmosphere boundary obtained from the amount of CO_2 absorbed by the alkali trap in a non-flow-through steady-state chamber
- F_c, CO_2 flux density at the soil-atmosphere boundary
- F_o, depth-integrated rate of subsurface CO_2 production expressed in units of flux density
- FT, flow-through
- NFT, non-flow-through
- NSS, non-steady-state
- SS, steady-state

NON-FLOW-THROUGH STEADY

STATE (NFT-SS) chambers—often called static chambers, absorption chambers, or alkali trap chambers—were the first devices used for measuring the flux of CO_2 from the soil surface (Bornemann, 1920; Lundegårdh, 1921) and were the only devices used very extensively for this purpose before the mid-1980s. Such chambers contain a trap filled with a known amount of alkali solution (or soda lime) that is supported above the soil surface, and they are typically deployed for long periods, often 12 or 24 h.

The amount of CO_2 trapped by the alkali is determined by titration (or by weight change for soda lime) and is usually corrected for CO_2 absorbed by an identical (blank) trap that is handled the same as other traps, but placed in a chamber deployed over a nonemitting surface. The resulting estimate of the soil respiration rate is often smaller than occasionally greater than estimates obtained from nonsteady-state (NSS) and flow-through steady-state (FT-SS) chamber systems, both of which gained rapid acceptance when accurate and portable CO_2 analysers became widely available about two decades ago. As a result, NFT-SS chamber estimates of soil respiration are considered by many investigators to be unreliable and, at best, to be estimates of the relative differences among various sources

The rate of gaseous CO_2 absorption by an alkali trap is proportional to the CO_2 concentration to which it is exposed. In a perfectly designed NFT-SS chamber, the absorption rate is equal to the surface flux, F_c, when headspace CO_2 is equal to the ambient level. In that situation, no change would occur in concentrations of the gas either above or below the soil surface during the period of deployment, F_c would remain equal to the underlying rate of CO_2 production, and the chamber would provide an accurate estimate of soil respiration.

However, achieving and maintaining such a balance is difficult, if not impossible, and the CO_2 concentration of both headspace and subsurface air often rises following chamber deployment until the rate of CO_2 absorption becomes equal to F_c The result is an underestimate of soil respiration—first, because the increase represents CO_2 production not accounted for in the alkali trap, and second, because the elevated concentrations support CO_2 losses by leakage through imperfect chamber seals and by lateral diffusion beneath the chamber walls. In contrast, headspace and subsurface CO_2 concentrations may decline following chamber deployment if F_c is particularly small. The result in this case is an overestimate of soil respiration, because the decline represents absorbed CO_2 not produced during the deployment period, and because the reduced concentrations support CO_2 gain by the chamber system via leakage and subsurface lateral diffusion.

The amount of increase (or decrease) in headspace CO_2 concentration depends not only on the magnitude of the soil respiration rate, but also on the efficiency of CO_2 absorption by the alkali trap, the effective diffusivity and storage coefficient of the gas in underlying soil, and its rate of exchange (leakage) between the chamber system and its surroundings. In a recent exhaustive review of existing literature regarding NFT-SS chambers, Rochette and Hutchinson (2003) concluded that:

- The optimal strength of the alkali solution is 0.5 to 1.0 *M*,
- The alkali trap should have total capacity approximately three times greater than the amount of CO_2 expected to be emitted during the deployment period,
- A 20% ratio of exposed alkali trap area to emitting soil surface area provides good absorption efficiency in many situations, but can be altered when needed to keep headspace CO_2 concentration as close as possible to the ambient level,
- The chamber should be nonvented and should have good seals that minimize CO_2 exchange between the chamber and its surroundings, and
- The deployment period should be at least 12 and preferably 24 h to minimize measurement bias due to the initial nonsteady-state condition, as well as bias due to chamber-induced temperature disturbances that often differ in algebraic sign between day and night (and thus tend to at least partially cancel across a 24-h deployment). Despite the decades of experience and scores of studies summarized by this review, however, many questions remain regarding the accuracy and functioning of NFT-SS chambers, as well as the optimal protocol for their use.

APPROACH

The transfer of gaseous CO_2 from its subsurface point of production to its carbonate form in the alkali trap of a NFT-SS chamber may be viewed as a seven-step process:

- Diffusive transport through air-filled soil pore space to the soil-atmosphere boundary,
- Diffusive transport through the interfacial layer of still air overlying the soil surface,
- Diffusive and/or convective transport from the top of that interfacial layer to the top of the interfacial layer overlying the alkali surface,
- Diffusive transport through this second interfacial layer of still air,
- Dissolution at the alkali solution surface,
- Chemical reaction with alkali near the solution surface, and
- Migration of dissolved carbonate away from the alkali surface.

There is considerable textbook and empirical evidence that the last three steps should be nonlimiting, and the weak dependence of absorption efficiency on alkali molarity greater than a modest threshold value supports this notion.

Because each of the first four steps is potentially rate-limiting, and all involve gaseous transport, we used a gas diffusion model to investigate their effect on CO_2 absorption efficiency, as well as their interaction with processes controlling CO_2 gain or loss by the chamber system. The model ignores diffusion of dissolved CO_2 and HCO^-_3 through soil water-filled pore space because of its small contribution to total transport. Other details regarding application of this model to chamber systems are available in Healy et al. and Hutchinson et al.

Briefly, we assumed that the soil was bare and had uniform properties with pH of 6.5, total porosity of 0.5, AFP of 0.3, tortuosity computed from the equation of Sallam et al. and effective diffusivity equal to the three-way product of tortuosity, AFP, and the binary molecular diffusion coefficient of CO_2 in air (*D*). Total porosity and AFP are understood to have units of $m^3\ m^{-3}$ soil.

Properties for CO_2 were chosen at 20°C and 100 kPa air pressure: D = 0.160 $cm^2\ s^{-1}$ ambient atmospheric concentration = 375 μmol mol^{-1}, Henry's Law coefficient for dissolution in water = 0.935 and the dissociation constant of dissolved CO_2 (H_2CO_3) = $10^{-6.40.}$ Equilibrium between gas phase and solution phase CO_2 and among dissolved species in soil water and in the alkali solution was assumed to occur instantaneously. The CO_2 was generated by a constant (or 24-h sinusoidal) zero-order source term with magnitude that decreased exponentially (with 10-cm relaxation depth) from a maximum at the soil surface to zero at the impermeable bottom of the simulated domain 50 cm below the surface.

The nonvented chamber headspace (30-cm diametre x 20-cm height) was assumed to be convectively mixed above a 0.5-cm diffusively mixed atmospheric interfacial layer, the depth of which was not altered by deployment of the chamber. Unless otherwise specified, the convection supported mixing equivalent to a tenfold increase in *D*. Chamber walls were assumed to be inserted 5 cm into the soil, except when studying the effect of changes in this parametre. Wall thickness was 0.25 cm except near the bottom, which we assumed was beveled to provide a cutting edge, as is often done to ease insertion and minimize the resulting potential for soil compaction. Finally, the cylindrical alkali container (5-cm internal depth x 0.2-cm walls) was mounted in the horizontal centre of the chamber with its bottom 2 cm above the soil surface. Its internal diametre was 13.4 cm when the ratio of exposed alkali surface area to emitting soil surface area was 0.20, and liquid depth (regardless of trap area) was 0.5 cm. The diffusively mixed atmospheric interfacial layer immediately above the alkali solution surface also had a 0.5-cm depth, and the convection in overlying air surrounded by vertical walls of the trap was assumed to support mixing equivalent to a threefold increase in *D* (compared with a tenfold increase in the bulk headspace).

Controllers of NFT-SS chamber performance that we investigated included the magnitude and diurnal periodicity of the soil respiration rate, the ratio of exposed alkali surface area to emitting soil surface area (hereafter abbreviated as the alkali:chamber area ratio), the efficiency of convective transport from the top of the soil interfacial layer to the top of the alkali interfacial layer, depth of the latter layer, height of the alkali trap above the soil surface, soil AFP, soil pH (which impacts the equilibrium between dissolved CO_2 and CO^-_3/HCO^-_3 in soil water), depth of chamber wall insertion into the soil, and the leakiness of chamber seals.

For convenience of the reader, the values assumed for these variables are listed at the top of all figures that report simulation results; the variable under examination in each case is identified by bold type.

Except in the simulations reported in the surface flux prior to and including the moment of chamber deployment was defined as the steady-state CO_2 flux density under the assumed conditions (equal to the depth-integrated rate of CO_2 production), which we designated F_o. Chamber performance was examined in the top graph of, 4, 5, and 6 by normalizing the instantaneous simulated flux into the chamber (F_c) with respect to F_o and plotting the result as a function of time after deployment. Deviations of the resulting trace from a horizontal straight line at $F_c/F_o = 1$ then represent a measure of chamber-induced perturbation of the pre-deployment CO_2 exchange rate.

The bottom graph in these three-panel shows the mean CO_2 concentration of the headspace as a function of time, while the bar graphs represent CO_2 accumulated by the alkali trap after several time periods (expressed as flux density, F_a); like F_c, F_a was normalized with respect to F_o. The solid curves in the top and bottom graphs of these figures identify the level at which the variable under investigation was held constant in other simulations while another was varied to study its influence on NFT-SS chamber performance.

Simulated non-flow-through steady-state chamber performance as a function of time when F_o (the depth-integrated rate of subsurface CO_2 production) followed a sinusoidal pattern with a 24-h period (solid curve in bottom panel graph); other simulation conditions are listed at the top of the figure. Curves with broken line types in the graph represent F_a (the alkali trap's instantaneous CO_2 absorption rate) during four simulated 24-h chamber deployment periods beginning at 0, 6, 12, or 18 h. Values in the centre-panel table are cumulative measurement error for the first 12 h or entire 24 h of each deployment period.

Simulated non-flow-through steady-state chamber performance as a function of time and F_o (the depth-integrated rate of subsurface CO_2 production) under conditions summarized above each three-panel array of graphs.

Chamber sidewalls were inserted (a) 5 cm or (b) 50 cm into the soil. The top and middle graphs show F_c (the surface flux of CO_2) and F_a (the alkali-measured flux of CO_2), respectively, both. normalized with respect to F_o; bottom graphs show mean headspace CO_2 concentration.

STEADY FLOW ENERGY EQUATION

The sketch above shows a piece of equipment such as a boiler, engine, pump, etc. through which fluid steadily flows. The fluid enters the equipment with velocity V_1 at an inlet 1 with area A_1 and leaves with velocity V_2 by an exhaust at 2 with area A_2. The heights of the inlet and exhaust above some reference datum are z_1 and z_2 respectively.

The equipment is enclosed within an imaginary surface called a control surface. Note that, unlike a system boundary, matter can cross a control surface. The control surface encloses a *control volume* and the objective here is to undertake an energy balance for that control volume.

Heat is transferred across the control surface at rate Q and work is being performed at rate by an output shaft. That work is called shaft work.

Consider a system that initially fills the control volume and a small section of the inlet plumbing outside the control volume. A small time $t\delta$later that system has moved so that it now coincides precisely with the control surface at the inlet, but part of the system has left the control volume at the exhaust.

STEADY FLOW PROCESSES

It should be clear from the above that the heat transfer in a steady flow process is the same as in the equivalent non-flow process, but the work w_s differs from the non-flow work because of the 'flow work' terms. For calculations one usually can find the heat transfer as for a non-flow process of the same kind and then find the shaft work w_s using the steady flow energy equation.

THERMODYNAMIC PARAMETERS

The central concept of thermodynamics is that of energy, the ability to do work. As stipulated by the first law, the total energy of the system and its surroundings is conserved. It may be transferred into a body by heating, compression, or addition of matter, and extracted from a body either by cooling, expansion, or extraction of matter.

For comparison, in mechanics, energy transfer results from a force which causes displacement, the product of the two being the amount of energy transferred. In a similar way, thermodynamic systems can be thought of as transferring energy as the result of a generalized force causing a generalized displacement, with the product of the two being the amount of energy transferred. These thermodynamic force-displacement pairs are known as conjugate variables. The most common conjugate thermodynamic variables are pressure-volume (mechanical parameters), temperature-entropy (thermal parameters), and chemical potential-particle number (material parameters).

THERMODYNAMIC INSTRUMENTS

There are two types of thermodynamic instruments, the meter and the reservoir. A thermodynamic meter is any device which measures any parameter of a thermodynamic system. In some cases, the thermodynamic parameter is actually defined in terms of an idealized measuring instrument. For example, the zeroth law states that if two bodies are in thermal equilibrium with a third body, they are also in thermal equilibrium with each other. This principle, as noted by James Maxwell in 1872, asserts that it is possible to measure temperature. An idealized thermometer is a sample of an ideal gas at constant pressure.

From the ideal gas law PV=nRT, the volume of such a sample can be used as an indicator of temperature; in this manner it defines temperature. Although pressure is defined mechanically, a pressure-measuring device, called a barometer may also be constructed from a sample of an ideal gas held at a constant temperature. A calorimeter is a device which is used to measure and define the internal energy of a system.

A thermodynamic reservoir is a system which is so large that it does not appreciably alter its state parameters when brought into contact with the test system. It is used to impose a particular value of a state parameter upon the system. For example, a pressure reservoir is a system at a particular pressure, which imposes that pressure upon any test system that it is mechanically connected to. The earth's atmosphere is often used as a pressure reservoir.

It is important that these two types of instruments are distinct. A meter does not perform its task accurately if it behaves like a reservoir of the state variable it is trying to measure. If, for example, a thermometer were to act as a temperature reservoir it would alter the temperature of the system being measured, and the reading would be incorrect. Ideal meters have no effect on the state variables of the system they are measuring.

THERMODYNAMIC STATES

When a system is at equilibrium under a given set of conditions, it is said to be in a definite state. The state of the system can be described by a number of intensive variables and extensive variables. The properties of the system can be described by an equation of state which specifies the relationship between these variables. State may be thought of as the instantaneous quantitative description of a system with a set number of variables held constant

Thermodynamics is an experimental science based on a small number of principles that are generalizations made from experience. It is concerned only with macroscopic or large-scale properties of matter and it makes no hypotheses about the small-scale or microscopic structure of matter. From the principles of thermodynamics one can derive general relations between such quantities as coefficients of expansion, compressibilities, specific heat capacities, heats of transformation, and magnetic and dielectric coefficients, especially as these are affected by temperature. The principles of thermodynamics also tell us which of these relations must be determined experimentally in order to completely specify all the properties of the system...

Thermodynamics is complementary to kinetic theory and statistical thermodynamics. Thermodynamics provides relationships between physical properties of any system once certain measurements are made. Kinetic theory and statistical thermodynamics enable one to calculate the mangitudes of these properties for those systems whose energy states can be determined.

CLASSICAL THERMODYNAMICS

We have learned that macroscopic quantities such as energy, temperature, and pressure are, in fact, statistical in nature: i.e., in equilibrium they exhibit random fluctuations about some mean value. If we were to plot out the probability distribution for the energy, say, of a system in thermal equilibrium with its surroundings we would obtain a Gaussian with a very small fractional width. In fact, we expect

$$\frac{\Delta^* E}{\bar{E}} \sim \frac{1}{\sqrt{f}},$$

where the number of degrees of freedom f is about 10^{24} for laboratory scale systems. This means that the statistical fluctuations of macroscopic quantities about their mean values are typically only about 1 in 10^{12}.

Since the statistical fluctuations of equilibrium quantities are so small, we can neglect them to an excellent approximation, and replace macroscopic quantities, such as energy, temperature, and pressure, by their mean values. So, $p \rightarrow \bar{p}$, and $T \rightarrow \bar{T}$, etc. In the following discussion, we shall drop the overbars altogether, so that p should be understood to represent the mean pressure $\bar{p}$, etc. This prescription, which is the essence of classical thermodynamics, is equivalent to replacing all statistically varying quantities by their most probable values.

Although there are formally four laws of thermodynamics (i.e., the zeroth to the third), the zeroth law is really a consequence of the second law, and the third law is actually only important at temperatures close to absolute zero. So, for most purposes, the two laws which really matter are the first law and the second law. For an infinitesimal process, the first law is written

$$đQ = dE + đW,$$

where dE is the change in internal energy of the system, $đQ$ is the heat absorbed by the system, and $đW$ is the work done by the system on its surroundings. Note that this is just a convention. We could equally well write the first law in terms of the heat emitted by the system or the work done on the system. It does not really matter, as long as we are consistent in our definitions.

The second law of thermodynamics implies that

$$đQ = TdS,$$

for a quasi-static process, where T is the thermodynamic temperature, and dS is the change in entropy of the system. Furthermore, for systems in which the only external parameter is the volume (i.e., gases), the work done on the environment is

$$đW = pdV,$$

where p is the pressure, and dV is the change in volume. Thus, it follows from the first and second laws of thermodynamics that

$$T\,dS = dE + p\,dV.$$

THE EQUATION OF STATE OF AN IDEAL GAS

Let us start our discussion by considering the simplest possible macroscopic system: i.e., an ideal gas. All of the thermodynamic properties of an ideal gas are summed up in its equation of state, which determines the relationship between its pressure, volume, and temperature. Unfortunately, classical thermodynamics is unable to tell us what this equation of state is from first principles. In fact, classical thermodynamics cannot tell us anything from first principles. We always have to provide some information to begin with before classical thermodynamics can generate any new results. This initial information may come from statistical physics (i.e., from our knowledge of the microscopic structure of the system under consideration), but, more usually, it is entirely empirical in nature (i.e., it is the result of experiments). Of course, the ideal gas law was first discovered empirically by Robert Boyle, but, nowadays, we can justify it from statistical arguments. Recall, that the number of accessible states of a monotonic ideal gas varies like

$$\Omega \propto V^N \chi(E),$$

where N is the number of atoms, and $\chi(E)$ depends only on the energy of the gas (and is independent of the volume). We obtained this result by integrating over the volume of accessible phase-space. Since the energy of an ideal gas is independent of the particle coordinates (because there are no interatomic forces), the integrals over the coordinates just reduced to N simultaneous volume integrals, giving the V^N factor in the above expression. The integrals over the particle momenta were more complicated, but were clearly completely independent of V, giving the $\chi(E)$ factor in the above expression. Now, we have a statistical rule which tells us that

$$X_a - \frac{1}{\beta}\frac{\partial \ln \Omega}{\partial x_a},$$

where X_a is the mean force conjugate to the external parameter x_a (i.e., $đW = \sum_a X_a dx_a$), and $\beta = 1/kT$. For an ideal gas, the only external parameter is the volume, and its conjugate force is the pressure (since $đW = pdV$). So, we can write

$$p = \frac{1}{\beta}\frac{\partial \ln \Omega}{\partial V}.$$

If we simply apply this rule to Eq. we obtain

$$p = \frac{N\,k\,T}{V}.$$

However, $N = v\,N_A$, where v is the number of moles, and N_A is Avagadro's number. Also, $k\,N_A = R$, where R is the ideal gas constant. This allows us to write the equation of state in its usual form

$$pV = vRT.$$

The above derivation of the ideal gas equation of state is rather elegant. It is certainly far easier to obtain the equation of state in this manner than to treat the atoms which make up the gas as little billiard balls which continually bounce of the walls of a container.

The latter derivation is difficult to perform correctly because it is necessary to average over all possible directions of atomic motion.

It is clear, from the above derivation, that the crucial element needed to obtain the ideal gas equation of state is the absence of interatomic forces. This automatically gives rise to a variation of the number of accessible states with E and V of the form which, in turn, implies the ideal gas law. So, the ideal gas law should also apply to polyatomic gases with no interatomic forces.

Polyatomic gases are more complicated that monatomic gases because the molecules can rotate and vibrate, giving rise to extra degrees of freedom, in addition to the translational degrees of freedom of a monatomic gas. In other words, $\chi(E)$, in Eq. becomes a lot more complicated in polyatomic gases. However, as long as there are no interatomic forces, the volume dependence of Ω is still V^N, and the ideal gas law should still hold true. In fact, we shall discover that the extra degrees of freedom of polyatomic gases manifest themselves by increasing the specific heat capacity.

There is one other conclusion we can draw from Eq. The statistical definition of temperature is

$$\frac{1}{kT} = \frac{\partial \ln \Omega}{\partial E}.$$

It follows that

$$\frac{1}{kT} = \frac{\partial \ln \chi}{\partial E}.$$

We can see that since χ is a function of the energy, but not the volume, then the temperature must be a function of the energy, but not the volume. We can turn this around and write

$$E = E(T).$$

In other words, the internal energy of an ideal gas depends only on the temperature of the gas, and is independent of the volume. This is pretty obvious, since if there are no interatomic forces then increasing the volume, which effectively increases the mean separation between molecules, is not going to affect the molecular energies in any way. Hence, the energy of the whole gas is unaffected.

The volume independence of the internal energy can also be obtained directly from the ideal gas equation of state. The internal energy of a gas can be considered as a general function of the temperature and volume, so

$$E = E(T, V).$$

It follows from mathematics that

$$dE = \left(\frac{\partial E}{\partial T}\right)_V dT + \left(\frac{\partial E}{\partial T}\right)_T dV,$$

where the subscript V reminds us that the first partial derivative is taken at constant volume, and the subscript T reminds us that the second partial derivative is taken at constant temperature. Thermodynamics tells us that for a quasi-static change of parameters

$$T\,dS = dE + p dV.$$

The ideal gas law can be used to express the pressure in term of the volume and the temperature in the above expression. Thus,

$$dS = \frac{1}{T} dE + \frac{\upsilon R}{V} dV.$$

Using Eq. this becomes

$$dS = \frac{1}{T}\left(\frac{\partial E}{\partial T}\right)_V dT + \left[\frac{1}{T}\left(\frac{\partial E}{\partial V}\right)_T + \frac{\upsilon R}{V}\right] dV.$$

However, dS is the exact differential of a well-defined state function, S. This means that we can consider the entropy to be a function of temperature and volume. Thus, $S = S(T, V)$, and mathematics immediately tells us that

$$dS = \left(\frac{\partial E}{\partial T}\right)_V dT + \left(\frac{\partial E}{\partial V}\right)_T dV.$$

The above expression is true for all small values of dT and dV, so a comparison with Eq. gives

$$\left(\frac{\partial S}{\partial T}\right)_V = \frac{1}{T}\left(\frac{\partial E}{\partial T}\right)_V,$$

$$\left(\frac{\partial S}{\partial T}\right)_T = \frac{1}{T}\left(\frac{\partial E}{\partial V}\right)_T + \frac{vR}{V}.$$

One well-known property of partial differentials is the equality of second derivatives, irrespective of the order of differentiation, so

$$\frac{\partial^2 S}{\partial V\,\partial T} = \frac{\partial^2 S}{\partial T\,\partial V}.$$

This implies that

$$\left(\frac{\partial}{\partial V}\right)_T\left(\frac{\partial S}{\partial T}\right)_V = \left(\frac{\partial}{\partial T}\right)_V\left(\frac{\partial S}{\partial T}\right)_T.$$

The above expression can be combined with Eqs. to give

$$\frac{1}{T}\left(\frac{\partial^2 E}{\partial V \partial T}\right) = \left[\frac{1}{T^2}\left(\frac{\partial E}{\partial V}\right)_T + \frac{1}{T}\left(\frac{\partial^2 E}{\partial T \partial V}\right)\right].$$

Since second derivatives are equivalent, irrespective of the order of differentiation, the above relation reduces to

$$\left(\frac{\partial E}{\partial V}\right)_T = 0,$$

which implies that the internal energy is independent of the volume for any gas obeying the ideal equation of state. This result was confirmed experimentally by James Joule in the middle of the nineteenth century.

HEAT CAPACITY OR SPECIFIC HEAT

Suppose that a body absorbs an amount of heat ΔQ and its temperature consequently rises by ΔT. The usual definition of the heat capacity, or specific heat, of the body is

$$C = \frac{\Delta Q}{\Delta T}.$$

If the body consists of v moles of some substance then the molar specific heat (i.e., the specific heat of one mole of this substance) is defined

$$c = \frac{1}{v}\frac{\Delta Q}{\Delta T}.$$

In writing the above expressions, we have tacitly assumed that the specific heat of a body is independent of its temperature. In general, this is not true. We can overcome this problem by only allowing the body in question to absorb a very small amount of heat, so that its temperature only rises slightly, and its specific heat remains approximately constant. In the limit as the amount of absorbed heat becomes infinitesimal, we obtain

$$c = \frac{1}{v}\frac{đW}{dT}.$$

In classical thermodynamics, it is usual to define two specific heats. Firstly, the molar specific heat at constant volume, denoted

$$c_p = \frac{1}{v}\left(\frac{đQ}{dT}\right)_V.$$

and, secondly, the molar specific heat at constant pressure, denoted

$$c_p = \frac{1}{v}\left(\frac{đQ}{dT}\right)_p.$$

Consider the molar specific heat at constant volume of an ideal gas. Since $dV = 0$, no work is done by the gas, and the first law of thermodynamics reduces to

$$đQ = dE.$$

It follows from Eq. that

$$c_V = \frac{1}{v}\left(\frac{\partial E}{\partial T}\right)_V.$$

Now, for an ideal gas the internal energy is volume independent. Thus, the above expression implies that the specific heat at constant volume is also volume independent. Since E is a function only of T, we can write

$$dE = \left(\frac{\partial E}{\partial T}\right)_V dT.$$

The previous two expressions can be combined to give

$$dE = v\, c_V\, dT$$

for an ideal gas.

Let us now consider the molar specific heat at constant pressure of an ideal gas. In general, if the pressure is kept constant then the volume changes, and so the gas does work on its environment. According to the first law of thermodynamics,

$$đQ = dE + pdV = vc_V\, dT + pdV.$$

The equation of state of an ideal gas tells us that if the volume changes by dV, the temperature changes by dT, and the pressure remains constant, then

$$pdV = vRdT.$$

The previous two equations can be combined to give

$$đQ = vVdT + vRdT.$$

Now, by definition

$$c_p = \frac{1}{v}\left(\frac{đQ}{dT}\right)_p,$$

so we obtain

$$c_p = c_V + R.$$

for an ideal gas. This is a very famous result. Note that at constant volume all of the heat absorbed by the gas goes into increasing its internal energy, and, hence, its temperature, whereas at constant pressure some of the absorbed heat is used to do work on the environment as the volume increases. This means that, in the latter case, less heat is available to increase the temperature of the gas. Thus, we expect the specific heat at constant pressure to exceed that at constant volume, as indicated by the above formula.

The ratio of the two specific heats Cp/C_V is conventionally denoted γ. We have

$$\gamma \equiv \frac{c_p}{c_V} = 1 + \frac{R}{c_V}.$$

for an ideal gas. In fact, γ is very easy to measure because the speed of sound in an ideal gas is written

$$c_s = \sqrt{\frac{\gamma p}{\rho}},$$

where $\rho$is the density. The extent of the agreement between g calculated from Eq. and the experimental g is quite remarkable.

Table. Specific Heats of Common Gases in Joules/mole/deg.
(at 15 C and 1 atm.) From Reif.3

Gas	*Symbol*	c_V	γ	γ
		(experiment)	*(experiment)*	*(theory)*
Helium	He	12.5	1.666	1.666
Argon	Ar	12.5	1.666	1.666
Nitrogen	N_2	20.6	1.405	1.407
Oxygen	O_2	21.1	1.396	1.397
Carbon Dioxide	CO_2	28.2	1.302	1.298
Ethane		39.3	1.22	1.214

CALCULATION OF SPECIFIC HEATS

Now that we know the relationship between the specific heats at constant volume and constant pressure for an ideal gas, it would be nice if we could calculate either one of these quantities from first principles. Classical thermodynamics cannot help us here. However, it is quite easy to calculate the specific heat at constant volume using our knowledge of statistical physics. Recall, that the variation of the number of accessible states of an ideal gas with energy and volume is written

$$\Omega(E,V) \propto V^N \chi(E).$$

For the specific case of a monatomic ideal gas, we worked out a more exact expression for Ω,

$$\Omega(E,V) = B\, V^N E^{3N/2},$$

where B is a constant independent of the energy and volume. It follows that

$$\ln\Omega = \ln B + N\ln V + \frac{3N}{2}\ln E.$$

The temperature is given by

$$\frac{1}{kT} = \frac{\partial \ln\Omega}{\partial E} = \frac{3N}{2}\frac{1}{E},$$

so

$$E = \frac{3}{2} N\, kT.$$

Since, $N = v\, N_A$, and $N_A\, k = R$, we can rewrite the above expression as

$$E = \frac{3}{2} v\, RT,$$

where R = 8.3143 joules/mole/deg, is the ideal gas constant. The above formula tells us exactly how the internal energy of a monatomic ideal gas depends on its temperature.

The molar specific heat at constant volume of a monatomic ideal gas is clearly

$$c_V = \frac{1}{V}\left(\frac{\partial E}{\partial T}\right)_v = \frac{3}{2} R.$$

This has the numerical value

$$c_V = 12.47 \text{ joules/mole/deg.}$$

Furthermore, we have

$$\text{cp} = \text{cV} + \text{R} = \frac{5}{2} R,$$

and

$$\gamma \equiv \frac{c_p}{c_V} = \frac{5}{3} = 1.667.$$

The predictions are borne out pretty well for the monatomic gases Helium and Argon. Note that the specific heats of polyatomic gases are larger than those of monatomic gases. This is because polyatomic molecules can rotate around their centres of mass, as well as translate, so polyatomic gases can store energy in the rotational, as well as the translational, energy states of their constituent particles.

ISOTHERMAL AND ADIABATIC EXPANSION

Suppose that the temperature of an ideal gas is held constant by keeping the gas in thermal contact with a heat reservoir. If the gas is allowed to expand quasi-statically under these so called isothermal conditions then the ideal equation of state tells us that

$$pV = \text{constant}.$$

This is usually called the isothermal gas law.

Suppose, now, that the gas is thermally isolated from its surroundings. If the gas is allowed to expand quasi-statically under these so called adiabatic conditions then it does work on its environment, and, hence, its internal energy is reduced, and its temperature changes. Let us work out the relationship between the pressure and volume of the gas during adiabatic expansion.

According to the first law of thermodynamics,

$$đQ = v\, c_V\, dT + pdV = 0,$$

in an adiabatic process (in which no heat is absorbed). The ideal gas equation of state can be differentiated, yielding

$$pdV + V\, dp = v\, RdT.$$

The temperature increment dT can be eliminated between the above two expressions to give

$$0 = \frac{c_V}{R}(pdV + V\, dp) + pdV = \left(\frac{c_V}{R} + 1\right) pdV + \frac{c_V}{R} V\, dp,$$

which reduces to

$$(c_V + R)p\, dV + c_V\, V\, dp = 0$$

Dividing through by $c_V\, p\, V$ yields

$$\gamma \frac{dV}{V} + \frac{dp}{p} = 0,$$

where

$$\gamma \equiv \frac{c_p}{c_V} + \frac{c_V + R}{c_V}.$$

It turns out that c_V is a very slowly varying function of temperature in most gases. So, it is always a fairly good approximation to treat the ratio of specific heats γ as a constant, at least over a limited temperature range. If γ is constant then we can integrate Eq. to give

$$\gamma \ln V + \ln p = \text{constant},$$

or

$$pV\,\gamma = \text{constant}$$

This is the famous adiabatic gas law. It is very easy to obtain similar relationships between V and T and P and T during adiabatic expansion or contraction. Since $p = vRT/V$, the above formula also implies that

$$\text{TV}^{-1} = \text{constant},$$

and

$$p^{1-\gamma}\, T^{\gamma} = \text{constant}$$

Equations are all completely equivalent.

TYPES OF THERMODYNAMIC PROCESSES

Paths through the space of thermodynamic variables are often specified by holding certain thermodynamic variables constant. It is useful to group these processes into pairs, in which each variable held constant is one member of a conjugate pair.

The pressure-volume conjugate pair is concerned with the transfer of mechanical or dynamic energy as the result of work.

- An isobaric process occurs at constant pressure. An example would be to have a movable piston in a cylinder, so that the pressure inside the cylinder is always at atmospheric pressure, although it is isolated from the atmosphere. In other words, the system is dynamically connected, by a movable boundary, to a constant-pressure reservoir.
- An isochoric (or isovolumetric) process is one in which the volume is held constant, meaning that the work done by the system will be zero. It follows that, for the simple system of two dimensions, any heat energy transferred to the system externally will be absorbed as internal energy. An isochoric process is also known as an isometric process. An example would be to place a closed tin can containing only air into a fire. To a first approximation, the can will not expand, and the only change will be that the gas gains internal energy, as evidenced by its increase in temperature and pressure. Mathematically, $\delta Q = dU$. We may say that the system is dynamically insulated, by a rigid boundary, from the environment

The temperature-entropy conjugate pair is concerned with the transfer of thermal energy as the result of heating.

- An isothermal process occurs at a constant temperature.
  An example would be to have a system immersed in a large constant-temperature bath. Any work energy performed by the system will be lost to the bath, but its temperature will remain constant. In other words, the system is thermally connected, by a thermally conductive boundary to a constant-temperature reservoir.

- An isentropic process occurs at a constant entropy. For a reversible process this is identical to an adiabatic process. If a system has an entropy which has not yet reached its maximum equilibrium value, a process of cooling may be required to maintain that value of entropy.
- An adiabatic process is a process in which there is no energy added or subtracted from the system by heating or cooling. For a reversible process, this is identical to an isentropic process. We may say that the system is thermally insulated from its environment and that its boundary is a thermal insulator. If a system has an entropy which has not yet reached its maximum equilibrium value, the entropy will increase even though the system is thermally insulated.

The above have all implicitly assumed that the boundaries are also impermeable to particles. We may assume boundaries that are both rigid and thermally insulating, but are permeable to one or more types of particle. Similar considerations then hold for the (chemical potential)-(particle number) conjugate pairs.

CONSERVATION OF ENERGY

In physics, the law of conservation of energy states that the total amount of energy in any isolated system remains constant but cannot be recreated, although it may change forms, e.g. friction turns kinetic energy into thermal energy. In thermodynamics, the first law of thermodynamics is a statement of the conservation of energy for thermodynamic systems, and is the more encompassing version of the conservation of energy. In short, the law of conservation of energy states that energy can not be created or destroyed, it can only be changed from one form to another or transferred from one body to another, but the total amount of energy remains constant (the same).

Ancient philosophers as far back as Thales of Miletus had inklings of the conservation of some underlying substance of which everything is made. However, there is no particular reason to identify this with what we know today as "mass-energy" (for example, Thales thought it was water) which can be described (in modern language) as conservatively converting potential energy to kinetic energy and back again. However, Galileo did not state the process in modern terms and again cannot be credited with the crucial insight.

It was Gottfried Wilhelm Leibniz during 1676–1689 who first attempted a mathematical formulation of the kind of energy which is connected with motion (kinetic energy). Leibniz noticed that in many mechanical systems (of several masses, mi each with velocity vi),

$$\sum_i m_i v_i^2$$

was conserved so long as the masses did not interact. He called this quantity the vis viva or living force of the system. The principle represents an accurate statement of the approximate conservation of kinetic energy in situations where there is no friction. Many physicists at that time were highly intelligent and held that the conservation of momentum, which holds even in systems with friction, as defined by the momentum:

$$\sum_i m_i v_i$$

was the conserved vis viva. It was later shown that, under the proper conditions, both quantities are conserved simultaneously such as in elastic collisions.

It was largely engineers such as John Smeaton, Peter Ewart, Karl Hotzmann, Gustave-Adolphe Hirn and Marc Seguin who objected that conservation of momentum alone was not adequate for practical calculation and who made use of Leibniz's principle.

The principle was also championed by some chemists such as William Hyde Wollaston. Academics such as John Playfair were quick to point out that kinetic energy is clearly not conserved. This is obvious to a modern analysis based on the second law of thermodynamics but in the 18th and 19th centuries, the fate of the lost energy was still unknown. Gradually it came to be suspected that the heat inevitably generated by motion under friction, was another form of vis viva.

In 1783, Antoine Lavoisier and Pierre-Simon Laplace reviewed the two competing theories of vis viva and caloric theory. Count Rumford's 1798 observations of heat generation during the boring of cannons added more weight to the view that mechanical motion could be converted into heat, and (as importantly) that the conversion was quantitative and could be predicted (allowing for a universal conversion constant between kinetic energy and heat). Vis viva now started to be known as energy, after the term was first used in that sense by Thomas Young in 1807.

The recalibration of vis viva to

$$\frac{1}{2}\sum_i m_i v_i^2$$

which can be understood as finding the exact value for the kinetic energy to work conversion constant, was largely the result of the work of Gaspard-Gustave Coriolis and Jean-Victor Poncelet over the period 1819–1839. The former called the quantity quantité de travail (quantity of work) and the latter, travail mécanique (mechanical work), and both championed its use in engineering calculation. It may appear, according to circumstances, as motion, chemical affinity, cohesion, electricity, light and magnetism; and from any one of these forms it can be transformed into any of the others."

A key stage in the development of the modern conservation principle was the demonstration of the mechanical equivalent of heat. The caloric theory maintained that heat could neither be created nor destroyed but conservation of energy entails the contrary principle that heat and mechanical work are interchangeable.

The mechanical equivalence principle was first stated in its modern form by the German surgeon Julius Robert von Mayer. Mayer reached his conclusion on a voyage to the Dutch East Indies, where he found that his patients' blood was a deeper red because they were consuming less oxygen, and therefore less energy, to maintain their body temperature in the hotter climate. He had discovered that heat and mechanical work were both forms of energy, and later, after improving his knowledge of physics, he calculated a quantitative relationship between them.

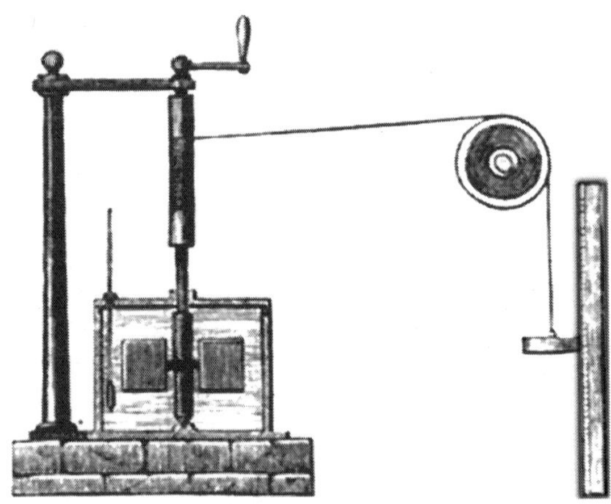

Fig. Joule's Apparatus for Measuring the Mechanical Equivalent of Heat.

Meanwhile, in 1843 James Prescott Joule independently discovered the mechanical equivalent in a series of experiments. In the most famous, now called the "Joule apparatus", a descending weight attached to a string caused a paddle immersed in water to rotate. He showed that the gravitational potential energy lost by the weight in descending was equal to the thermal energy (heat) gained by the water by friction with the paddle.

Over the period 1840–1843, similar work was carried out by engineer Ludwig A. Colding though it was little known outside his native Denmark.

Both Joule's and Mayer's work suffered from resistance and neglect but it was Joule's that, perhaps unjustly, eventually drew the wider recognition.In 1844, William Robert Grove postulated a relationship between mechanics, heat, light, electricity and magnetism by treating them all as manifestations of a single "force" (energy in modern terms).

DEFINITION OF THERMODYNAMIC STATE

A thermodynamic state is a condition of a system, which is explained by its parameters. The state of any system can be described by its parameters, such as temperature, pressure, density and composition, which are independent of surroundings. The thermodynamic state of a system is defined by its values and properties for all fluid systems.

Example, the state of an electric battery requires amount of electric charge it contains.

Properties are extensive or intensive. Extensive properties are additive. Thus, if the system is divided into a number of sub-systems, the value of the property for whole system is equal to sum of its values for the parts. *Volume* is an extensive property. Intensive properties do not depend on the quantity of matter present. *Temperature* and *pressure* are intensive properties. Specific properties are extensive properties per unit mass and are denoted by lower case letters.

Eg:

Specific Volume = V / m = n

Specific properties are intensive because they do not depend on mass of the system. The properties of a simple system are uniform throughout. In general, the properties of a system vary from point to point. We can usually study a general system by sub-dividing it into a number of simple systems in each of which the properties are uniform.

Equations of Thermodynamic State

It is a fact that two properties are needed to define the state of any pure substance in equilibrium or quasi-steady process.

Thus for a simple gas,

$$P = P(\nu, T), \text{ or } \nu = \nu(P, T), \text{ or } T = T(P, \nu),$$

where ν is the volume per unit mass 1 / p

By putting values of ν and T we get value of P.

From the equation f (P, v, T) = 0, the equation of state for an ideal gas for applications are:

$$P\overline{\nu} = RT$$

where

$\overline{\nu}$ is the volume per mol of gas

R is the "Universal Gas Constant 8.31 KJ / Kmol - K .

A form of this equation which is more useful in fluid flow problems is obtained if we divide by the molecular weight, M :

$$P\nu = RT, \text{ or } P = pRT$$

where

R is R / M

R = 0.287 KJ / Kg – K

BOUNDARIES

A boundary is a real or imaginary demarcated region, which is drawn around a thermodynamic, across which the quantities such as heat, mass, or work can flow.

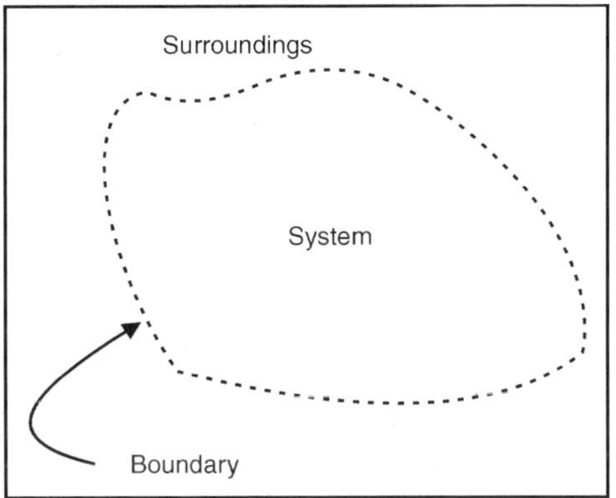

It is a division between a system and its surroundings. A boundary may be adiabatic, isothermal, diathermal, insulating, permeable, or semi permeable.

Boundaries can be fixed such as a constant volume reactor or moveable piston as shown in figure 1.5

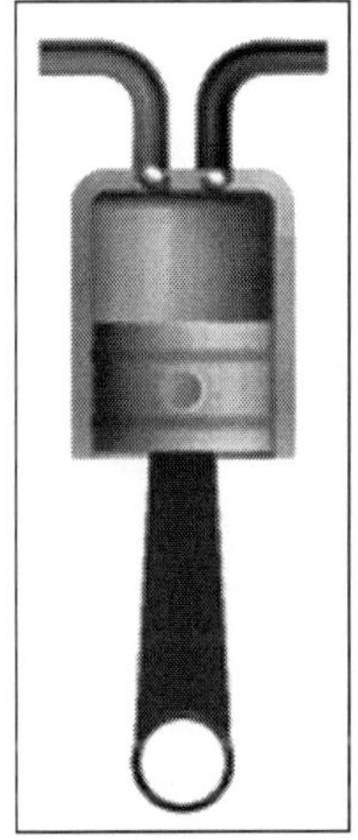

We can also say that a "boundary" is simply an imaginary dotted line drawn around the volume of something in which there is going to be a change in the internal energy. Anything that passes across the boundary effects a change in the internal energy.

For an engine, a fixed boundary means the piston, which is locked at its position such as constant volume process. In that engine, a moveable boundary allows the piston to move in and out. The closed systems, boundaries are real while for open system boundaries are imaginary.

SURROUNDINGS

Surrounding are the Material or regions in space, which are outside the boundaries of the system, which interact and influence system behavior.

THERMODYNAMICS SYSTEM

A thermodynamic system is that part of the universe that is under consideration. A real or imaginary boundary separates the system from the rest of the universe, which is called as an environment.

The diagram of thermodynamics system is shown below:

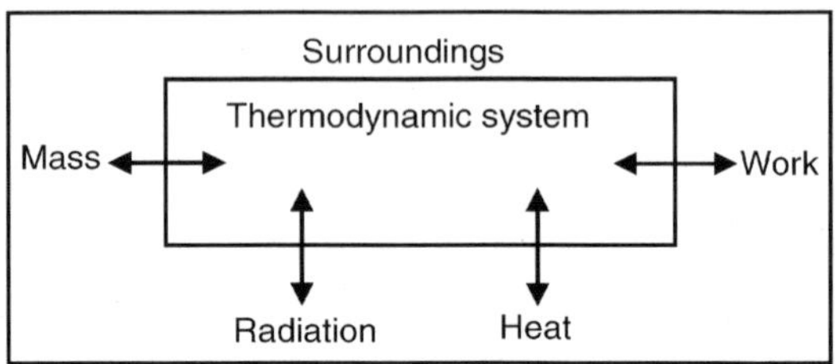

The classification of thermodynamic systems is based on the nature of its boundary and the flows of matter, energy and entropy.

ISOLATED SYSTEM

An isolated system is that which does not exchange heat, matter or work with an environment.

An example of an isolated system is insulated gas cylinder. A system can never be fully isolated from its environment because there is always some slight coupling with minimum gravitational attraction.

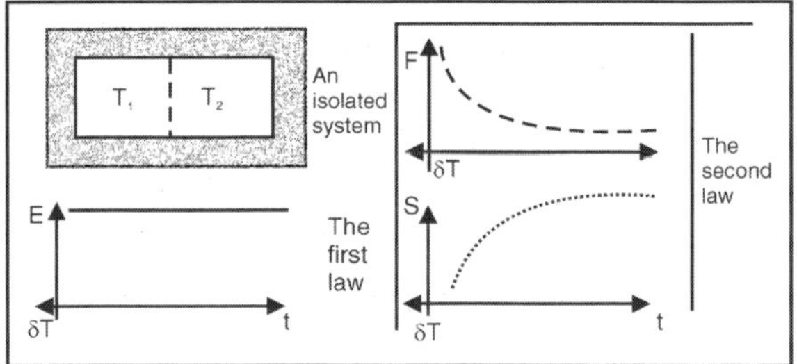

Figure shows the essence of thermodynamics.

In the above system, heat within a gas of temperature T_2, will flow in time t, toward a gas of temperature T_1, where $T_2 > T_1$ and $T = T_2 - T_1$, thus conserving the system's total energy E, free energy F, decreases the entropy S, rises as T 0 at equilibrium.

CLOSED SYSTEM

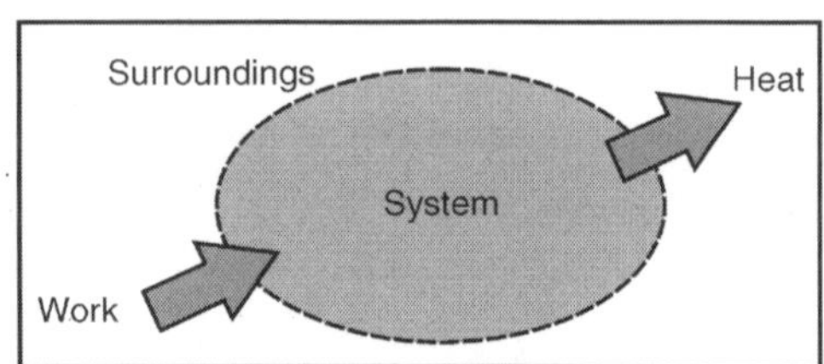

A Closed system is that which exchanges energy and not the matter with the environment. A greenhouse is an example of a closed system.

Another example is heat engine shown in Figure below.

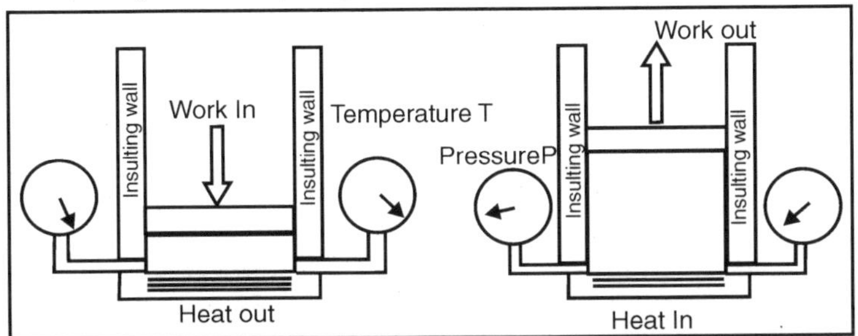

A Heat engine is a device, which converts heat energy into mechanical energy. As work is done on the gas, the temperature and pressure increase and heat will be transferred out of the system. When heat is transferred to the system, the gas expands, which does the work on its surroundings..

Example: *Piston and cylinder - a closed system*

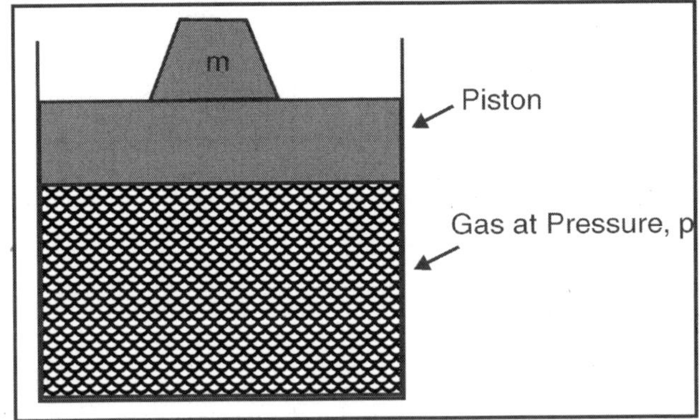

OPEN SYSTEM

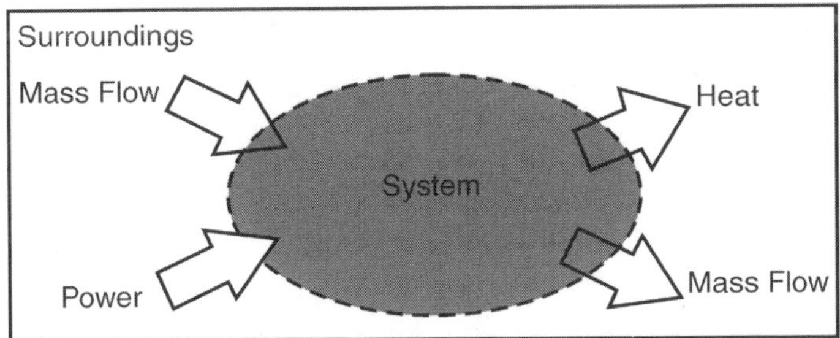

An Open System is that in which the exchange of energy and matter exists with environment. In this a boundary allows the matter to exchange, which is called as permeable. It's possible for an open system to import order and export disorder.

Consider an open system as shown in Figure below.

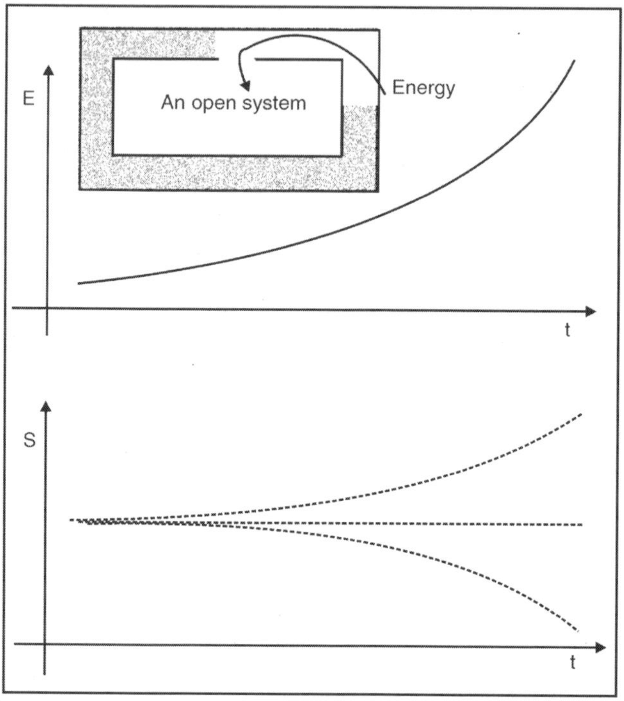

In the above system, energy can enter the system from outside environment, which increases the system's total energy E, over the time t. Such energy flow can lead to an increase, decrease or net change at all in entropy S, of a system.

Example: *The gas turbine engine*

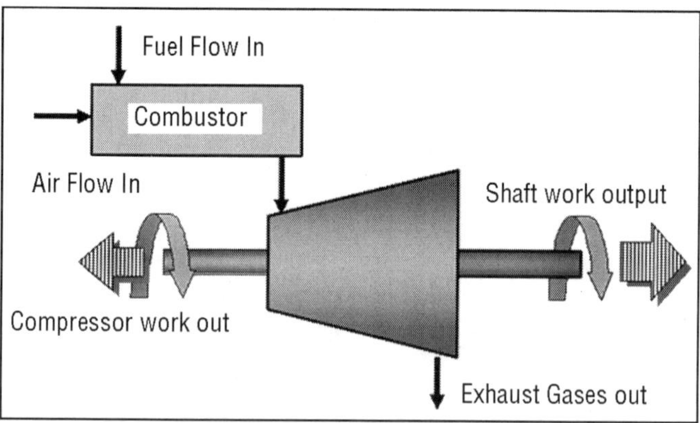

PURE SUBSTANCES AND SIMPLE COMPRESSIBLE SUBSTANCES

If a substance has a homogeneous chemical composition, even though it may consist of more than one phase, it is said to be a*pure substance*, provided, of course, that each phase has the same chemical composition. Thus, water is a pure substance even when it appears as solid, liquid, and vapor simultaneously. In many applications, air may be taken to be a pure substance even though it is a mixture of several species. If the composition doesn't change during the process under consideration, the fact that air is really a mixture can be ignored safely. Often this results in a great savings in computation. The analyst must use judgment in determining when a composite substance may be considered pure.

In many processes, surface effects, electrical effects, magnetic effects, elastic effects, etc. are not important. In this case, the only form of work that will be considered can be computed from changes in pressure and volume. Under these conditions, the substance is said to be a *simple compressible substance* and the expressions for derived properties in terms of fundamental properties are especially simple as we shall see in the section on the First Law. Remember, as in much of thermodynamics, whether or not a substance can be classified as a simple compressible substance depends upon the circumstances as well as the substance.

CONTROL VOLUME

A *control volume* is a fixed region in space, which is used for study of mass and energy, to balance the systems. The boundary of the control volume may be real or imaginary. The *control surface* is the boundary of the control volume.

ADIABATIC SYSTEM

The adiabatic system is a process during which the heat contents of the system remain constant. In this there is no flow of heat across the boundaries of system. Such process is called as adiabatic process or system.

HOMOGENEOUS SYSTEM

The system that has single or uniform phase such as like solid, liquid or gas is called as homogeneous system.

A system is said to be homogeneous if the physical states of all matter are uniform. eg. Mixture of gases.

HETEROGENEOUS SYSTEM

The system that has more than one phase such as combination of solid, liquid and gaseous state is called as heterogeneous system.

A system is said to be heterogeneous, if its contents does not possess the same physical state.

For eg:

Immiscible liquids

Solid in contact with an immiscible liquid

Solid in contact with a gas.

4

The Equipartition Theorem

INTERNAL ENERGY OF A MONATOMIC IDEAL GAS

The internal energy of a monatomic ideal gas containing *N* particles. This means that each particle possess, on average, units of energy. Monatomic particles have only three translational degrees of freedom, corresponding to their motion in three dimensions. They possess no internal rotational or vibrational degrees of freedom. Thus, the mean energy per degree of freedom in a monatomic ideal gas is. In fact, this is a special case of a rather general result. Let us now try to prove this. Suppose that the energy of a system is determined by some generalized coordinates and corresponding generalized momenta *Pk*, so that

$$E = E(q_1, ..., q_f, p_1, ..., p_f).$$

Suppose further that:

The total energy splits additively into the form

$$E = \epsilon_i (p_i) + E'(q_1, ..., p_f),$$

where ϵ_i involves only one variable p_i, and the remaining part E' does not depend on p_i.

The function ϵ_i is quadratic in p_i, so that

$$\epsilon_i (p_i) + bp_i^2,$$

where *b* is a constant.

The most common situation in which the above assumptions are valid is where p_i is a momentum. This is because the kinetic energy is usually a quadratic function of each momentum component, whereas the potential energy does not involve the momenta at all. However, if a coordinate q_i were to satisfy assumptions 1 and 2 then the theorem we are about to establish would hold just as well.

What is the mean value of ϵ_i in thermal equilibrium if conditions 1 and 2 are satisfied? If the system is in equilibrium at absolute temperature $T \equiv (k\beta)^{-1}$ then it is distributed according to the Boltzmann distribution. In the classical approximation, the mean value of $\in_i$ is expressed in terms of integrals over all phase-space:

$$\overline{\in}_i = \frac{\int_{-\infty}^{\infty} \exp[-\beta E(q_1,...,p_f)] \in_i dq_1...dp_f}{\int_{-\infty}^{\infty} \exp[-\beta E(q_1,...,p_f)] dq_1...dp_f}.$$

Condition 1 gives

$$\overline{\in}_i = \frac{\int_{-\infty}^{\infty} \exp[-\beta(\in_i + E')] \in_i dq_1...dp_f}{\int_{-\infty}^{\infty} \exp[-\beta(\in_i + E')] dq_1...dp_f}$$

$$= \frac{\int_{-\infty}^{\infty} \exp(-\beta \in_i) \in_i dp_i \int_{-\infty}^{\infty} \exp(-\beta E') dq_1...dp_f}{\int_{-\infty}^{\infty} \exp(-\beta \in_i) dp_i \int_{-\infty}^{\infty} \exp(-\beta E') dq_1...dp_f},$$

where use has been made of the multiplicative property of the exponential function, and where the last integrals in both the numerator and denominator extend over all variables *qk*and *pk* except p_i. These integrals are equal and, thus, cancel. Hence,

$$\overline{\in}_i = \frac{\int_{-\infty}^{\infty} \exp(-\beta \in_i) \in_i dq_i}{\int_{-\infty}^{\infty} \exp(-\beta \in_i) dp_i}.$$

This expression can be simplified further since

$$\int_{-\infty}^{\infty} \exp(-\beta \in_i) \in_i dp_i \equiv -\frac{\theta}{\partial \beta}\left[\int_{-\infty}^{\infty} \exp(-\beta \in_i) dp_i\right],$$

so

$$\overline{\in}_i = -\frac{\partial}{\partial \beta} \ln\left[\int_{-\infty}^{\infty} \exp(-\beta \in_i) dp_i\right].$$

According to condition 2,

where $y = \sqrt{\beta}\ p_i$. Thus,

$$\ln \int_{-\infty}^{\infty} \exp(-\beta \in_i) dp_i = -\frac{1}{2} \ln \beta + \ln \int_{-\infty}^{\infty} \exp(-by^2) dy.$$

Note that the integral on the right-hand side does not depend on β at all. It follows from equation that

$$\overline{\in}_i = -\frac{\partial}{\partial \beta}\left(-\frac{1}{2} \ln \beta\right) = \frac{1}{2\beta},$$

giving

$$\overline{\epsilon}_i = \frac{1}{2}kT.$$

This is the famous equipartition theorem of classical physics. It states that the mean value of every independent quadratic term in the energy is equal to (1/2)kT. If all terms in the energy are quadratic then the mean energy is spread equally over all degrees of freedom (hence the name "equipartition").

THE CLASSICAL APPROXIMATION OF EQUIPARTITION THEOREM

HARMONIC OSCILLATORS

Our proof of the equipartition theorem depends crucially on the classical approximation. To see how quantum effects modify this result, let us examine a particularly simple system which we know how to analyze using both classical and quantum physics: i.e., a simple harmonic oscillator. Consider a one-dimensional harmonic oscillator in equilibrium with a heat reservoir at temperature T. The energy of the oscillator is given by

$$E = \frac{p^2}{2m} + \frac{1}{2}kx^2,$$

where the first term on the right-hand side is the kinetic energy, involving the momentum p and mass m, and the second term is the potential energy, involving the displacement x and the force constant k. Each of these terms is quadratic in the respective variable. So, in the classical approximation the equipartition theorem yields:

$$\frac{\overline{p^2}}{2m} = \frac{1}{2}kT,$$

$$\frac{1}{2}k\overline{x^2} = \frac{1}{2}kT.$$

That is, the mean kinetic energy of the oscillator is equal to the mean potential energy which equals (1/2)kT. It follows that the mean total energy is

$$\overline{E} = \frac{1}{2}kT + \frac{1}{2}kT = kT.$$

According to quantum mechanics, the energy levels of a harmonic oscillator are equally spaced and satisfy

$$E_n = (n + 1/2)\,\hbar\omega,$$

where n is a non-negative integer, and

$$\omega = \sqrt{\frac{k}{m}}.$$

The partition function for such an oscillator is given by

$$Z = \sum_{n=0}^{\infty} \exp(-\beta E_n) = \exp[-(1/2)\beta\hbar\omega] \sum_{n=0}^{\infty} \exp(-n\beta\hbar\omega).$$

Now,

$$\sum_{n=0}^{\infty} \exp(-n\beta\hbar\omega) = 1 + \exp(-\beta\hbar\omega) + \exp(-2\,\beta\hbar\omega) + \cdots$$

is simply the sum of an infinite geometric series, and can be evaluated immediately,

$$\sum_{n=0}^{\infty} \exp(-n\beta\hbar\omega) = \frac{1}{1-\exp(-\beta\hbar\omega)}.$$

Thus, the partition function takes the form

$$Z = \frac{\exp[-(1/2)\beta\hbar\omega]}{1-\exp(-\beta\hbar\omega)},$$

and

$$\ln Z = -\frac{1}{2}\beta\hbar\omega - \ln[1-\exp(-\beta\hbar\omega)].$$

The mean energy of the oscillator is given by equation

$$\bar{E} = -\frac{\partial}{\partial\beta}\ln Z = -\left[-\frac{1}{2}\hbar\omega - \frac{\exp(-\beta\hbar\omega)\,\hbar\omega}{1-\exp(-\beta\hbar\omega)}\right],$$

or

$$\bar{E} = \hbar\omega\left[-\frac{1}{2} + \frac{1}{\exp(\beta\hbar\omega)-1}\right].$$

Consider the limit

$$\beta\hbar\omega = \frac{\hbar\omega}{kT} \ll 1,$$

in which the thermal energy kT is large compared to the separation $\hbar\omega$ between the energy levels. In this limit,

$$\exp(\beta\hbar\omega) \simeq 1 + \beta\hbar\omega,$$

so

$$\bar{E} \simeq \hbar\omega\left[\frac{1}{2} + \frac{1}{\beta\hbar\omega}\right] \simeq \hbar\omega\left[\frac{1}{\beta\hbar\omega}\right],$$

giving

$$\bar{E} \simeq \frac{1}{\beta} = kT.$$

Thus, the classical result holds whenever the thermal energy greatly exceeds the typical spacing between quantum energy levels.

Consider the limit

$$\beta\hbar\omega\frac{\hbar\omega}{kT} >> 1,$$

in which the thermal energy is small compared to the separation between the energy levels. In this limit,

$$\exp(\beta\hbar\omega) >> 1,$$

Thus, if the thermal energy is much less than the spacing between quantum states then the mean energy approaches that of the ground-state (the so-called zero point energy). Clearly, the equipartition theorem is only valid in the former limit, where $kT >> \hbar\omega$, and the oscillator possess sufficient thermal energy to explore many of its possible quantum states.

ENTROPIES OF SUBSTANCES

SPECIFIC HEATS

We have discussed the internal energies and entropies of substances (mostly ideal gases) at some length. Unfortunately, these quantities cannot be directly measured. Instead, they must be inferred from other information. The thermodynamic property of substances which is the easiest to measure is, of course, the heat capacity, or specific heat. In fact, once the variation of the specific heat with temperature is known, both the internal energy and entropy can be easily reconstructed via

$$E(T, V) = v\int_0^T c_V(T,V)dT + E(0,V),$$

$$S(T, V) =$$

Here, use has been made of $dS = đQ/T$, and the third law of thermodynamics. Clearly, the optimum way of verifying the results of statistical thermodynamics is to compare the theoretically predicted heat capacities with the experimentally measured values.

Classical physics, in the guise of the equipartition theorem, says that each independent degree of freedom associated with a quadratic term in the energy possesses an average energy (1/2)kT in thermal equilibrium at temperature T. Consider a substance made up of N molecules. Every molecular degree of freedom contributes (1/2)$N\ k\ T$, or (1/2) vRT, to the mean energy of the substance (with the tacit proviso that each degree of freedom is associated with a quadratic term in the energy). Thus, the contribution to the molar heat capacity at constant volume (we wish to avoid the complications associated with any external work done on the substance) is

$$\frac{1}{v}\left(\frac{\partial E}{\partial T}\right)_V = \frac{1}{v}\frac{\partial[(1/2)vRT]}{\partial T} = \frac{1}{2}R,$$

per molecular degree of freedom. The total classical heat capacity is therefore

$$c_V = \frac{g}{2} R,$$

where g is the number of molecular degrees of freedom. Since large complicated molecules clearly have very many more degrees of freedom than small simple molecules, the above formula predicts that the molar heat capacities of substances made up of the former type of molecules should greatly exceed those of substances made up of the latter. In fact, the experimental heat capacities of substances containing complicated molecules are generally greater than those of substances containing simple molecules, but by nowhere near the large factor predicted by previous equation.

This equation also implies that heat capacities are temperature independent. In fact, this is not the case for most substances. Experimental heat capacities generally increase with increasing temperature. These two experimental facts pose severe problems for classical physics. Incidentally, these problems were fully appreciated as far back as 1850. Stories that physicists at the end of the nineteenth century thought that classical physics explained absolutely everything are largely apocryphal.

The equipartition theorem (and the whole classical approximation) is only valid when the typical thermal energy kT greatly exceeds the spacing between quantum energy levels. Suppose that the temperature is sufficiently low that this condition is not satisfied for one particular molecular degree of freedom. In fact, suppose that kT is much less than the spacing between the energy levels.

In this situation the degree of freedom only contributes the ground-state energy, E_0, say, to the mean energy of the molecule. The ground-state energy can be a quite complicated function of the internal properties of the molecule, but is certainly not a function of the temperature, since this is a collective property of all molecules. It follows that the contribution to the molar heat capacity is

$$\frac{1}{v}\left(\frac{\partial[N\, E_0]}{\partial T}\right)_v = 0.$$

Thus, if kT is much less than the spacing between the energy levels then the degree of freedom contributes nothing at all to the molar heat capacity. We say that this particular degree of freedom is frozen out. Clearly, at very low temperatures just about all degrees of freedom are frozen out. As the temperature is gradually increased, degrees of freedom successively "kick in," and eventually contribute their full (1/2) R to the molar heat capacity, as kT approaches, and then greatly exceeds, the spacing between their quantum energy levels. We can use these simple ideas to explain the behaviours of most experimental heat capacities. To make further progress, we need to estimate the typical spacing between the quantum energy levels associated with various degrees of freedom. We can do this by observing the frequency of the electromagnetic radiation emitted and absorbed during transitions between these energy levels. If the typical spacing between energy levels is ΔE then transitions between the various levels are associated with photons of frequency v, where $hv = \Delta E$.

We can define an effective temperature of the radiation via $hv = kT_{rad}$. If $T \gg T_{rad}$ then $kT >> \Delta E$, and the degree of freedom makes its full contribution to the heat capacity. On the other hand, if $T \ll T_{rad}$ then $kT << \Delta E$, and the degree of freedom is frozen out. The "temperatures" of various different types of radiation. It is clear that degrees of freedom which give rise to emission or absorption of radio or microwave radiation contribute their full (1/2)R to the molar heat capacity at room temperature.

Degrees of freedom which give rise to emission or absorption in the visible, ultraviolet, X–ray, or γ–ray regions of the electromagnetic spectrum are frozen out at room temperature. Degrees of freedom which emit or absorb infrared radiation are on the border line.

Table. Effective "Temperatures" of Various Types of Electromagnetic Radiation

Radiation type	Frequency (Hz)	T_{rad} (°K)
Radio	$< 10^9$	< 0.05
Microwave	$10^9 - 10^{11}$	< 0.05
Infrared	$10^{11} - 10^{14}$	5. 5000
Visible	5×10^{14}	2×10^4
Ultraviolet	$10^{15} - 10^{17}$	5×10^4 5×10^5
X-ray	$10^{17} - 10^{20}$	5×10^6 5×10^9
g-ray	$> 10^{20}$	$> 5 \times 10^93$

INVESTIGATE THE SPECIFIC HEATS OF GASES

Specific Heats of Gases

Let us now investigate the specific heats of gases. Consider, first of all, translational degrees of freedom. Every molecule in a gas is free to move in three dimensions. If one particular molecule has mass m and momentum $p = mv$ then its kinetic energy of translation is

$$K = \frac{1}{2m}\left(p_x^2 + p_y^2 + p_z^2\right).$$

The kinetic energy of other molecules does not involve the momentum P of this particular molecule. Moreover, the potential energy of interaction between molecules depends only on their position coordinates, and, thus, certainly does not involve P. Any internal rotational, vibrational, electronic, or nuclear degrees of freedom of the molecule also do not involve P.

Hence, the essential conditions of the equipartition theorem are satisfied (at least, in the classical approximation). Since equation contains three independent quadratic terms, there are clearly three degrees of freedom associated with translation (one for each dimension of space), so the translational contribution to the molar heat capacity of gases is

$$(c_V)_{\text{translation}} = \frac{3}{2}R.$$

Suppose that our gas is contained in a cubic enclosure of dimensions L. According to Schrödinger's equation, the quantized translational energy levels of an individual molecule are given by

$$E = \frac{\hbar^2\pi^2}{2mL^2}\left(n_1^2 + n_2^2 + n_3^2\right),$$

where n_1, n_2, and n_3 are positive integer quantum numbers. Clearly, the spacing between the energy levels can be made arbitrarily small by increasing the size of the enclosure. This implies that translational degrees of freedom can be treated classically, so that equation is always valid (except very close to absolute zero). We conclude that all gases possess a minimum molar heat capacity of (3/2) R due to the translational degrees of freedom of their constituent molecules.

The electronic degrees of freedom of gas molecules (i.e., the possible configurations of electrons orbiting the atomic nuclei) typically give rise to absorption and emission in the ultraviolet or visible regions of the spectrum. It follows from electronic degrees of freedom are frozen out at room temperature.

Similarly, nuclear degrees of freedom (i.e., the possible configurations of protons and neutrons in the atomic nuclei) are frozen out because they are associated with absorption and emission in the X–ray and γ–ray regions of the electromagnetic spectrum. In fact, the only additional degrees of freedom we need worry about for gases are rotational and vibrational degrees of freedom. These typically give rise to absorption lines in the infrared region of the spectrum.

The rotational kinetic energy of a molecule tumbling in space can be written

$$K = \frac{1}{2}I_x\omega_x^2 + \frac{1}{2}I_y\omega_y^2 + \frac{1}{2}I_z\omega_z^2,$$

where the x –, y–, and z–axes are the so called principle axes of inertia of the molecule (these are mutually perpendicular), ω_x, ω_y, and ω_z are the angular velocities of rotation about these axes, and I_x, I_y, and I_z are the moments of inertia of the molecule about these axes. No other degrees of freedom depend on the angular velocities of rotation.

Since the kinetic energy of rotation is the sum of three quadratic terms, the rotational contribution to the molar heat capacity of gases is

$$(c_V)_{\text{rotation}} = \frac{3}{2}R,$$

according to the equipartition theorem. Note that the typical magnitude of a molecular moment of inertia is md^2, where m is the molecular mass, and d is the typical interatomic spacing in the molecule. A special case arises if the molecule is linear (e.g. if the molecule is diatomic).

In this case, one of the principle axes lies along the line of centers of the atoms. The moment of inertia about this axis is of order ma^2, where a is a typical nuclear dimension (remember that nearly all of the mass of an atom resides in the nucleus). Since $a \sim 10^{-5}\, d$,it follows that the moment of inertia about the line of centres is minuscule compared to the moments of inertia about the other two principle axes. In quantum mechanics, angular momentum is quantized in units of $\hbar$. The energy levels of a rigid rotator are written

$$E = \frac{\hbar^2}{2I} J(J+1),$$

where I is the moment of inertia and J is an integer. Note the inverse dependence of the spacing between energy levels on the moment of inertia.

It is clear that for the case of a linear molecule, the rotational degree of freedom associated with spinning along the line of centres of the atoms is frozen out at room temperature, given the very small moment of inertia along this axis, and, hence, the very widely spaced rotational energy levels.

Classically, the vibrational degrees of freedom of a molecule are studied by standard normal mode analysis of the molecular structure. Each normal mode behaves like an independent harmonic oscillator, and, therefore, contributes R to the molar specific heat of the gas [(1/2) R from the kinetic energy of vibration and (1/2)R from the potential energy of vibration].

A molecule containing n atoms has $n - 1$ normal modes of vibration. For instance, a diatomic molecule has just one normal mode (corresponding to periodic stretching of the bond between the two atoms). Thus, the classical contribution to the specific heat from vibrational degrees of freedom is

$$(c_V)_{\text{vibration}} = (n-1)R.$$

The rotational and vibrational degrees of freedom actually make a contribution to the specific heats of gases at room temperature, once quantum effects are taken into consideration? We can answer this question by examining just one piece of data. The infrared absorption spectrum of Hydrogen Chloride. The absorption lines correspond to simultaneous transitions between different vibrational and rotational energy levels.

Hence, this is usually called a vibration-rotation spectrum. The missing line at about 3.47 microns corresponds to a pure vibrational transition from the ground-state to the first excited state (pure vibrational transitions are forbidden: HCl molecules always have to simultaneously change their rotational energy level if they are to couple effectively to electromagnetic radiation).

The longer wavelength absorption lines correspond to vibrational transitions in which there is a simultaneous decrease in the rotational energy level. Likewise, the shorter wavelength absorption lines correspond to vibrational transitions in which there is a simultaneous increase in the rotational energy level.

It is clear that the rotational energy levels are more closely spaced than the vibrational energy levels. The pure vibrational transition gives rise to absorption at about 3.47 microns, which corresponds to infrared radiation of frequency 8.5×10^{11} hertz with an associated radiation "temperature" of 4400 degrees kelvin.

We conclude that the vibrational degrees of freedom of HCl, or any other small molecule, are frozen out at room temperature. The rotational transitions split the vibrational lines by about 0.2 microns.

This implies that pure rotational transitions would be associated with infrared radiation of frequency 5×10^{12} hertz and corresponding radiation "temperature" 260 degrees kelvin. We conclude that the rotational degrees of freedom of HCl, or any other small molecule, are not frozen out at room temperature, and probably contribute the classical $(1/2)R$ to the molar specific heat. There is one proviso, however. Linear molecules (like HCl) effectively only have two rotational degrees of freedom (instead of the usual three), because of the very small moment of inertia of such molecules along the line of centres of the atoms.

We are now in a position to make some predictions regarding the specific heats of various gases. Monatomic molecules only possess three translational degrees of freedom, so monatomic gases should have a molar heat capacity $(3/2)R = 12.47$ joules/degree/mole.

The ratio of specific heats $\gamma = Cp/C_V = (C_V + R)/C_V$ should be $5/3 = 1.667$. It can be seen from both of these predictions are borne out pretty well for Helium and Argon. Diatomic molecules possess three translational degrees of freedom and two rotational degrees of freedom (all other degrees of freedom are frozen out at room temperature). Thus, diatomic gases should have a molar heat capacity $(5/2)R = 20.8$ joules/degree/mole.

The ratio of specific heats should be $7/5 = 1.4$ It can be seen from these are pretty accurate predictions for Nitrogen and Oxygen. The freezing out of vibrational degrees of freedom becomes gradually less effective as molecules become heavier and more complex.

This is partly because such molecules are generally less stable, so the force constant k is reduced, and partly because the molecular mass is increased. Both these effect reduce the frequency of vibration of the molecular normal modes and, hence, the spacing between vibrational energy levels.

This accounts for the obviously non-classical [i.e., not a multiple of $(1/2)R$] specific heats of Carbon Dioxide and Ethane. In both molecules, vibrational degrees of freedom contribute to the molar specific heat (but not the full R because the temperature is not high enough). The variation of the molar heat capacity at constant volume (in units of R) of gaseous hydrogen with temperature. The expected contribution from the translational degrees of freedom is $(3/2)R$ (there are three translational degrees of freedom per molecule).

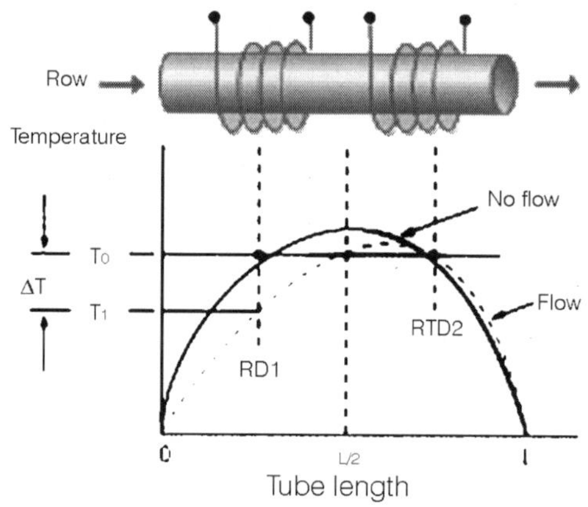

Fig. The Molar Heat Capacity at Constant Volume.

The expected contribution at high temperatures from the rotational degrees of freedom is R (there are effectively two rotational degrees of freedom per molecule). Finally, the expected contribution at high temperatures from the vibrational degrees of freedom is R (there is one vibrational degree of freedom per molecule). It can be seen that as the temperature rises the rotational, and then the vibrational, degrees of freedom eventually make their full classical contributions to the heat capacity.

AMPLITUDE VIBRATIONS

Specific Heats of Solids

Consider a simple solid containing N atoms. Now, atoms in solids cannot translate (unlike those in gases), but are free to vibrate about their equilibrium positions. Such vibrations are called lattice vibrations, and can be thought of as sound waves propagating through the crystal lattice. Each atom is specified by three independent position coordinates, and three conjugate momentum coordinates. Let us only consider small amplitude vibrations. In this case, we can expand the potential energy of interaction between the atoms to give an expression which is quadratic in the atomic displacements from their equilibrium positions. It is always possible to perform a normal mode analysis of the oscillations. In effect, we can find $3N$ independent modes of oscillation of the solid.

Each mode has its own particular oscillation frequency, and its own particular pattern of atomic displacements. Any general oscillation can be written as a linear combination of these normal modes. Let q_i be the (appropriately normalized) amplitude of the ith normal mode, and p_i the momentum conjugate to this coordinate. In normal mode coordinates, the total energy of the lattice vibrations takes the particularly simple form

$$E = \frac{1}{2}\sum_{i=1}^{3N}(p_i^2 + \omega_i^2 q_i^2),$$

where ω_i is the (angular) oscillation frequency of the ith normal mode. It is clear that in normal mode coordinates, the linearized lattice vibrations are equivalent to $3N$ independent harmonic oscillators (of course, each oscillator corresponds to a different normal mode).

The typical value of ω_i is the (angular) frequency of a sound wave propagating through the lattice. Sound wave frequencies are far lower than the typical vibration frequencies of gaseous molecules. In the latter case, the mass involved in the vibration is simply that of the molecule, whereas in the former case the mass involved is that of very many atoms (since lattice vibrations are non-localized).

The strength of interatomic bonds in gaseous molecules is similar to those in solids, so we can use the estimate $\omega \sim \sqrt{k/m}$ (k is the force constant which measures the strength of interatomic bonds, and m is the mass involved in the oscillation) as proof that the typical frequencies of lattice vibrations are very much less than the vibration frequencies of simple molecules.

It follows from $\Delta E = \hbar\omega$ that the quantum energy levels of lattice vibrations are far more closely spaced than the vibrational energy levels of gaseous molecules. Thus, it is likely (and is, indeed, the case) that lattice vibrations are not frozen out at room temperature, but, instead, make their full classical contribution to the molar specific heat of the solid.

If the lattice vibrations behave classically then, according to the equipartition theorem, each normal mode of oscillation has an associated mean energy kT in equilibrium at temperature T [$(1/2)kT$ resides in the kinetic energy of the oscillation, and $(1/2)kT$ resides in the potential energy]. Thus, the mean internal energy per mole of the solid is

$$\overline{E} = 3NkT = 3\upsilon RT.$$

It follows that the molar heat capacity at constant volume is

$$c_V = \frac{1}{\upsilon}\left(\frac{\partial \overline{E}}{\partial T}\right)_V = 3R$$

for solids. This gives a value of 24.9 joules/mole/degree. In fact, at room temperature most solids (in particular, metals) have heat capacities which lie remarkably close to this value. This fact was discovered experimentally by Dulong and Petite at the beginning of the nineteenth century, and was used to make some of the first crude estimates of the molecular weights of solids (if we know the molar heat capacity of a substance then we can easily work out how much of it corresponds to one mole, and by weighing this amount, and then dividing the result by Avogadro's number, we can obtain an estimate of the molecular weight).

The experimental molar heat capacities *Cp* at constant pressure for various solids. The heat capacity at constant volume is somewhat less than the constant pressure value, but not by much, because solids are fairly incompressible. It can be seen that Dulong and Petite's law (i.e., that all solids have a molar heat capacities close to 24.9 joules/mole/degree) holds pretty well for metals. However, the law fails badly for diamond. This is not surprising.

As is well-known, diamond is an extremely hard substance, so its intermolecular bonds must be very strong, suggesting that the force constant k is large. Diamond is also a fairly low density substance, so the mass m involved in lattice vibrations is comparatively small. Both these facts suggest that the typical lattice vibration frequency of diamond ($\omega \sim \sqrt{k/m}$) is high. In fact, the spacing between the different vibration energy levels (which scales like $\hbar\omega$) is sufficiently large in diamond for the vibrational degrees of freedom to be largely frozen out at room temperature.

This accounts for the anomalously low heat capacity of diamond.

Table. Values of (Joules/Mole/Degree)
For Some Solids at T = 298° K. K. From Reif.

Solid	Cp	Solid	Cp
Copper	24.5	Aluminium	24.4
Silver	25.5	Tin (white)	26.4
Lead	26.4	Sulphur (rhombic)	22.4
Zinc	25.4	Carbon (diamond)	6.1

Dulong and Petite's law is essentially a high temperature limit. The molar heat capacity cannot remain a constant as the temperature approaches absolute zero, since, by equation this would imply $S \to \infty$, which violates the third law of thermodynamics. We can make a crude model of the behaviour of C_V at low temperatures by assuming that all the normal modes oscillate at the same frequency, ω, say.

According to equation the solid acts like a set of $3N$ independent oscillators which, making use of Einstein's approximation, all vibrate at the same frequency. We can use the quantum mechanical result for a single oscillator to write the mean energy of the solid in the form

$$\bar{E} = 3N\hbar\omega\left(\frac{1}{2} + \frac{1}{\exp(\beta\hbar\omega) - 1}\right).$$

The molar heat capacity is defined

$$c_V = \frac{1}{v}\left(\frac{\partial \bar{E}}{\partial T}\right)_V = \frac{1}{v}\left(\frac{\partial \bar{E}}{\partial T}\right)_V \frac{\partial \beta}{\partial T} = -\frac{1}{vkT^2}\left(\frac{\partial \bar{E}}{\partial T}\right)_V,$$

giving

$$c_V = -\frac{3\,N_A\hbar\omega}{kT^2}\left[-\frac{\exp(\beta\hbar\omega)}{[\exp(\beta\hbar\omega)-1]^2}\right],$$

which reduces to

$$c_V = 3R\left(\frac{\theta_E}{T}\right)^2\frac{\exp(\theta_E/T)}{[\exp(\theta_E/T)-1]^2}.$$

Here,

$$\theta_E = \frac{\hbar\omega}{k}$$

is called the Einstein temperature. If the temperature is sufficiently high that T>> θ_E then $kT >> \hbar\omega$, and the above expression reduces to $C_V = 3R$,, after expansion of the exponential functions. Thus, the law of Dulong and Petite is recovered for temperatures significantly in excess of the Einstein temperature. On the other hand, if the temperature is sufficiently low that T<< θ_E then the exponential factors in equation become very much larger than unity, giving

$$C_V \sim 3R\left(\frac{\theta_E}{T}\right)\exp(-\theta_E/T).$$

So, in this simple model the specific heat approaches zero exponentially as $T \to 0$. In reality, the specific heats of solids do not approach zero quite as quickly as suggested by Einstein's model when $T \to 0$. The experimentally observed low temperature behaviour is more like $C_V \propto T^3$. The reason for this discrepancy is the crude approximation that all normal modes have the same frequency. In fact, long wavelength modes have lower frequencies than short wavelength modes, so the former are much harder to freeze out than the latter (because the spacing between quantum energy levels, $\hbar\omega$, is smaller in the former case).

The molar heat capacity does not decrease with temperature as rapidly as suggested by Einstein's model because these long wavelength modes are able to make a significant contribution to the heat capacity even at very low temperatures. A more realistic model of lattice vibrations was developed by the Dutch physicist Peter Debye in 1912. In the Debye model, the frequencies of the normal modes of vibration are estimated by treating the solid as an isotropic continuous medium. This approach is reasonable because the only modes which really matter at low temperatures are the long wavelength modes: i.e., those whose wavelengths greatly exceed the interatomic spacing. It is plausible that these modes are not particularly sensitive to the discrete nature of the solid: i.e., the fact that it is made up of atoms rather than being continuous.

Consider a sound wave propagating through an isotropic continuous medium. The disturbance varies with position vector r and time t like exp[$-i$

(k.r – ωt)], where the wave-vector k and the frequency of oscillation ω satisfy the dispersion relation for sound waves in an isotropic medium:

$$\omega = kC_s.$$

Here, C_S is the speed of sound in the medium. Suppose, for the sake of argument, that the medium is periodic in the x–, y–, and z–directions with periodicity lengths L_x, L_y, and L_z, respectively. In order to maintain periodicity we need

$$k_x(x + L_x) = k_x + 2\pi n_x,$$

where n_x is an integer. There are analogous constraints on k_y and k_z. It follows that in a periodic medium the components of the wave-vector are quantized, and can only take the values

$$k_x = \frac{2\pi}{L_x} n_x,$$

$$k_y = \frac{2\pi}{L_y} n_y,$$

$$k_z = \frac{2\pi}{L_z} n_z,$$

where n_x, n_y, and n_z are all integers. It is assumed that L_x, L_y, and L_z are macroscopic lengths, so the allowed values of the components of the wave-vector are very closely spaced. For given values of k_y and k_z, the number of allowed values of k_x which lie in the range k_x to $k_x + dk_x$ is given by

$$\Delta n_x = \frac{L_x}{2\pi} 2k_x.$$

It follows that the number of allowed values of k (i.e., the number of allowed modes) when k_x lies in the range k_z to $k_z + dk_z$, k_z, + k_y lies in the range k_y to $k_y + dk_y$, and k_z lies in the range k_z to $k_z + dk_z$, is

$$\rho d^3k = \left(\frac{L_x}{2\pi} dk_x\right)\left(\frac{L_y}{2\pi} dk_y\right)\left(\frac{L_z}{2\pi} dk_z\right) = \frac{V}{(2\pi)^3} dk_x dk_y dk_z,$$

where $V = L_x L_y L_z$ is the periodicity volume, and $d^3k \equiv dk_x dk_y dk_z$. The quantity ρ is called the density of modes. Note that this density is independent of k, and proportional to the periodicity volume.

Thus, the density of modes per unit volume is a constant independent of the magnitude or shape of the periodicity volume. The density of modes per unit volume when the magnitude of k lies in the range k to $k + dk$ is given by multiplying the density of modes per unit volume by the "volume" in k-space of the spherical shell lying between radii k and $k + dk$. Thus,

$$\rho_k\, dk = \frac{4\pi k^2\, dk}{(2\pi)} = \frac{V^2}{2\pi^2} dk.$$

Consider an isotropic continuous medium of volume V. According to the above relation, the number of normal modes whose frequencies lie between ω and $\omega + d\omega$ (which is equivalent to the number of modes whose k values lie in the range ω/C_s to $\omega/C_s + d\omega/C_s$) is

$$\sigma_C(\omega)\, d\omega = 3\frac{k^2\, V}{2\pi^2} dk = 3\frac{V^2}{2\pi^2 {C_S}^3}\omega^2 d\omega.$$

The factor of 3 comes from the three possible polarizations of sound waves in solids. For every allowed wavenumber (or frequency) there are two independent torsional modes, where the displacement is perpendicular to the direction of propagation, and one longitudinal mode, where the displacement is parallel to the direction of propagation. Torsion waves are vaguely analogous to electromagnetic waves (these also have two independent polarizations). The longitudinal mode is very similar to the compressional sound wave in gases. Of course, torsion waves can not propagate in gases because gases have no resistance to deformation without change of volume.

The Debye approach consists in approximating the actual density of normal modes $\sigma(\omega)$ by the density in a continuous medium $\sigma_C(\omega)$, not only at low frequencies (long wavelengths) where these should be nearly the same, but also at higher frequencies where they may differ substantially.

Suppose that we are dealing with a solid consisting of N atoms. We know that there are only $3N$ independent normal modes. It follows that we must cut off the density of states above some critical frequency, ω_D say, otherwise we will have too many modes. Thus, in the Debye approximation the density of normal modes takes the form

$$\sigma_D(\omega) = \sigma_C(\omega) \text{ for } \omega < \omega_D$$
$$\sigma_D(\omega) = 0 \text{ for } \omega < \omega_D$$

Here, ω_D is the Debye frequency. This critical frequency is chosen such that the total number of normal modes is $3N$, so

$$\int_0^\infty \sigma_C(\omega)\, d\omega = \int_0^{\omega_D} \sigma_C(\omega)\, d\omega = 3N.$$

Substituting equation into the previous formula yields

$$\frac{3V}{2\pi^2 C_s^3}\int_0^{\omega_D} d\omega = \frac{3V}{2\pi^2 C_s^3}\omega_D^3 = 3N.$$

This implies that

$$\omega_D = C_s\left(6\pi^2\frac{N}{V}\right)^{1/3}.$$

Thus, the Debye frequency depends only on the sound velocity in the solid and the number of atoms per unit volume. The wavelength corresponding to the Debye frequency is $2\pi\, C_s/\omega_D$, which is clearly on the order of the interatomic spacing $a \sim (V/N)^{1/3}$.

It follows that the cut-off of normal modes whose frequencies exceed the Debye frequency is equivalent to a cut-off of normal modes whose wavelengths are less than the interatomic spacing. Of course, it makes physical sense that such modes should be absent.

Compares the actual density of normal modes in diamond with the density predicted by Debye theory. Not surprisingly, there is not a particularly strong resemblance between these two curves, since Debye theory is highly idealized. Nevertheless, both curves exhibit sharp cut-offs at high frequencies, and coincide at low frequencies. Furthermore, the areas under both curves are the same. This is sufficient to allow Debye theory to correctly account for the temperature variation of the specific heat of solids at low temperatures.

We can use the quantum mechanical expression for the mean energy of a single oscillator, equation to calculate the mean energy of lattice vibrations in the Debye approximation. We obtain

$$\overline{E} = \int_0^\infty \sigma_D(\omega)\hbar\omega\left(\frac{1}{2} + \frac{1}{\exp(\beta\hbar\omega) - 1}\right)d\omega.$$

According to equation the molar heat capacity takes the form

$$C_V = \frac{1}{vkT^2}\int_0^\infty \sigma_D(\omega)\hbar\omega\left[\frac{\exp(\beta\hbar\omega)\hbar\omega}{[\exp(\beta\hbar\omega) - 1]^2}\right]d\omega.$$

Substituting in equation we find that

$$C_V = \frac{k}{v}\int_0^{\omega_D} \frac{\exp(\beta\hbar\omega)(\beta\hbar\omega)^2}{[\exp(\beta\hbar\omega) - 1]^2}\frac{3V}{2\pi^2 C_s^3}\omega^2 d\omega,$$

giving

$$C_V = \frac{3Vk}{2\pi^2 v(C_s\beta\hbar)^3}\int_0^{\beta\hbar\omega_D} \frac{\exp x}{(\exp x - 1)^2}x^4 dx,$$

in terms of the dimensionless variable $x = \beta\hbar\omega$. According to equation the volume can be written

$$V = 6\pi^2 N\left(\frac{Cs}{\omega_D}\right)^3,$$

so the heat capacity reduces to

$$C_V = 3Rf_D(\beta\hbar\omega_D) = 3Rf_D(\theta_D / T),$$

where the Debye function is defined

$$f_D(y) \equiv \frac{3}{y^3}\int_0^y \frac{\exp x}{(\exp x - 1)^2} x^4 dx.$$

We have also defined the Debye temperature θ_D as

$$k\theta_D = \hbar\omega_D.$$

Consider the asymptotic limit in which $T \gg \theta_D$. For small y, we can approximate exp x as $1 + x$ in the integrand of equation so that

$$f_D(y) \to \frac{3}{y^3}\int_0^y x^2 dx = 1.$$

Thus, if the temperature greatly exceeds the Debye temperature we recover the law of Dulong and Petite that $C_V = 3R$. Consider, now, the asymptotic limit in which $T \ll \theta_D$. For large y,

$$\int_0^y \frac{\exp x}{(\exp x - 1)^2} x^4 dx \simeq \int_0^\infty \frac{\exp x}{(\exp x - 1)^2} x^4 dx = \frac{4\pi^4}{15}.$$

Thus, in the low temperature limit

$$f_D(y) \to \frac{4\pi^4}{5}\frac{1}{y^3}.$$

This yields

$$c_V \simeq \frac{12\pi^4}{5} R\left(\frac{T}{\theta_D}\right)^3$$

in the limit $T \ll \theta_D$: i.e., c_V varies with temperature like T^3.

Table. Comparison of Debye Temperatures (In Degrees Kelvin)

Solid	θ_D from low temp.	θ_D from sound speed
NaCl	308	320
KCl	230	246
Ag	225	216
Zn	308	305

The fact that c_V goes like T^3 at low temperatures is quite well verified experimentally, although it is sometimes necessary to go to temperatures as low as 0.02 θ_D to obtain this asymptotic behaviour.

Theoretically, θ_D should be calculable from equation in terms of the sound speed in the solid and the molar volume. A comparison of Debye temperatures evaluated by this means with temperatures obtained empirically by fitting the law to the low temperature variation of the heat capacity. It can be seen that there is fairly good agreement between the theoretical and empirical Debye temperatures. This suggests that the Debye theory affords a good, thought not perfect, representation of the behaviour of c_V in solids over the entire temperature range.

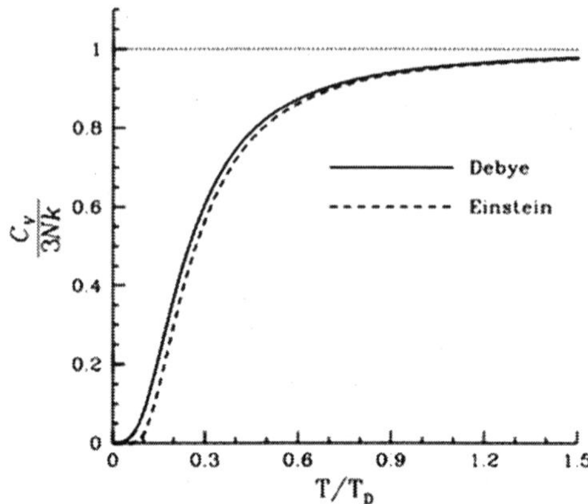

Fig. The Molar Heat Capacity of Various Solids.

Finally, the actual temperature variation of the molar heat capacities of various solids as well as that predicted by Debye's theory. The prediction of Einstein's theory is also show for the sake of comparison. Note that 24.9 joules/mole/degree is about 6 calories/gram-atom/degree (the latter are chemist's units).

5

Entropy

INTRODUCTION

The definition of the absolute scale of temperature, as given in relations that in passing at constant temperature 0 from one adiabatic 4' to any other adiabatic 0", the quotient H/o of the heat absorbed by the temperature at which it is absorbed is the same for the same two adiabatics whatever the temperature of the isothermal path. This quotient is called the change of entropy. In passing along an adiabatic there is no change of entropy, since no heat is absorbed. The adiabatics are lines of constant entropy, and are also called Isentropics.

In virtue of relations the change of entropy of a substance between any two states depends only on the initial and final states, and may be reckoned along any reversible path, not necessarily isothermal, by dividing each small increment of heat, dH, by the temperature, 0, at which it is acquired, and taking the sum or integral of the quotients, dH/o, so obtained. In thermodynamics entropy, symbolized by S, is a measure of the unavailability of a system's energy to do work.It is a measure of the randomness of molecules in a system and is central to the second law of thermodynamics and the fundamental thermodynamic relation, which deal with physical processes and whether they occur spontaneously. Spontaneous changes, in isolated systems, occur with an increase in entropy.

Spontaneous changes tend to smooth out differences in temperature, pressure, density, and chemical potential that may exist in a system, and entropy is thus a measure of how far this smoothing-out process has progressed. When a system's energy is defined as the sum of its "useful" energy, and its "useless energy", i.e. that energy which cannot be used for external work, then entropy may be visualized as the "scrap" or "useless" energy whose energetic prevalence over the total energy of a system is directly proportional to the absolute temperature of the considered system..

Entropy is a function of a quantity of heat which shows the possibility of conversion of that heat into work. The increase in entropy is small when heat is added at high temperature and is greater when heat is added at lower temperature. Thus for maximum entropy there is minimum availability for conversion into work and for minimum entropy there is maximum availability for conversion into work. Entropy S is not defined directly, but rather by an equation relating the change in entropy of the system to the change in heat of the system.

For a constant temperature, the change in entropy ΔS is defined by the equation ΔS = ΔQ/T, where ΔQ is the amount of heat absorbed in an isothermal and reversible process in which the system goes from one state to another, and T is the absolute temperature at which the process is occurring. If the temperature of the system is not constant, then the relationship becomes a differential equation dS = dQ/T. To understand what this equation means, suppose the temperature T can be expressed as a function T(Q) of the heat Q. Then the total change in entropy as the heat-level varies is:

$$\Delta S = \int_A \frac{1}{T(Q)} dQ$$

where A is the set defining the range of heat values in the system.

Entropy is one of the factors that determines the free energy of the system. This thermodynamic definition of entropy is only valid for a system in equilibrium (because temperature is defined only for a system in equilibrium), while the statistical definition of entropy applies to any system. Thus the statistical definition is usually considered the fundamental definition of entropy. Entropy increase has often been defined as a change to a more disordered state at a molecular level. In recent years, entropy has been interpreted in terms of the "dispersal" of energy. Entropy is an extensive state function that accounts for the effects of irreversibility in thermodynamic systems.In terms of statistical mechanics, the entropy describes the number of the possible microscopic configurations of the system. The statistical definition of entropy is the more fundamental definition, from which all other definitions and all properties of entropy follow.

In a thermodynamic system, a "universe" consisting of "surroundings" and "systems" and made up of quantities of matter, its pressure differences, density differences, and temperature differences all tend to equalize over time – simply because equilibrium state has higher probability than any other. In the ice melting example, the difference in temperature between a warm room (the surroundings) and cold glass of ice and water (the system and not part of the room), begins to be equalized as portions of the heat energy from the warm surroundings spread out to the cooler system of ice and water. Over time the temperature of the glass and its contents and the temperature of the room become equal. The entropy of the room has decreased as some of its energy has been dispersed to the ice and water. However, as calculated in the example, the entropy of the system of ice and water has increased more than the entropy of the surrounding room has decreased. In an isolated system such as the room and ice water taken together, the dispersal of energy from warmer to cooler always results in a net increase in entropy.

Thus, when the 'universe' of the room and ice water system has reached a temperature equilibrium, the entropy change from the initial state is at a maximum. The entropy of the thermodynamic system is a measure of how far the equalization has progressed. A special case of entropy increase, the entropy of mixing, occurs when two or more different substances are mixed. If the substances are at the same temperature and pressure, there will be no net exchange of heat or work – the entropy increase will be entirely due to the mixing of the different substances.

From a macroscopic perspective, in classical thermodynamics the entropy is interpreted simply as a state function of a thermodynamic system: that is, a property depending only on the current state of the system, independent of how that state came to be achieved. The state function has the important property that, when multiplied by a reference temperature, it can be understood as a measure of the amount of energy in a physical system that cannot be used to do thermodynamic work; i.e., work mediated by thermal energy.

More precisely, in any process where the system gives up energy ΔE, and its entropy falls by ΔS, a quantity at least TR ΔS of that energy must be given up to the system's surroundings as unusable heat (TR is the temperature of the system's external surroundings). Otherwise the process will not go forward.

Quantitatively, Clausius states the mathematical expression for this theorem is as follows. Let δQ be an element of the heat given up by the body to any reservoir of heat during its own changes, heat which it may absorb from a reservoir being here reckoned as negative, and T the absolute temperature of the body at the moment of giving up this heat, then the equation:

$$\int \frac{\delta Q}{T} = 0$$

must be true for every reversible cyclical process, and the relation:

$$\int \frac{\delta Q}{T} \geq 0$$

must hold good for every cyclical process which is in any way possible. This is the essential formulation of the second law and one of the original forms of the concept of entropy. The dimensions of entropy are energy divided by temperature, which is the same as the dimensions of Boltzmann's constant (kB) and heat capacity. The SI unit of entropy is "joule per kelvin" (J·K"1). In this manner, the quantity "ΔS" is utilized as a type of internal energy, which accounts for the effects of irreversibility, in the energy balance equation for any given system. In the Gibbs free energy equation, i.e. ΔG = ΔH – TΔS, for example, which is a formula commonly utilized to determine if chemical reactions will occur, the energy related to entropy changes TΔS is subtracted from the "total" system energy ΔH to give the "free" energy ΔG of the system, as during a chemical process or as when a system changes state.

CHANGE IN ENTROPY

Entropy is used in everyday life as a synonym for chaos. For example: the entropy in my room increases as the semester goes on. But it's also been used to describe the approach to an imagined final state of the universe when everything reaches the same temperature: the entropy is supposed to increase to a maximum, then nothing will ever happen again.

Of course, the first law—the conservation of total energy including heat energy—is easy to express quantitatively: one only needs to find the equivalence factor between heat units and energy units, calories to joules, since all the other types of energy are already in joules, add it all up to get the total and that will remain constant.

The second law, that heat only flows from a warmer body to a colder one, does have quantitative consequences: the efficiency of any reversible engine has to equal that of the Carnot cycle, and any nonreversible engine has less efficiency.

Entropy (S) is very difficult to visualize because it does not represent anything tangible. The entropy increase δS is the heat transfer to a substance δQ divided by the absolute temperature of the substance (T) during a reversible heat-transfer process.

At the very simplest level, on a plot of Absolute Temperature (T) against Entropy (Q/T = S) for a reversible cyclic process (as shown below) the area enclosed = Q

Note: The reversible cyclic process shown below is actually the theoretical Carnot cycle.

1->2 being isothermal expansion.
2->3 being adiabetic expansion.
3->4 being isothermal compression.
4->1 being adiabetic compression.

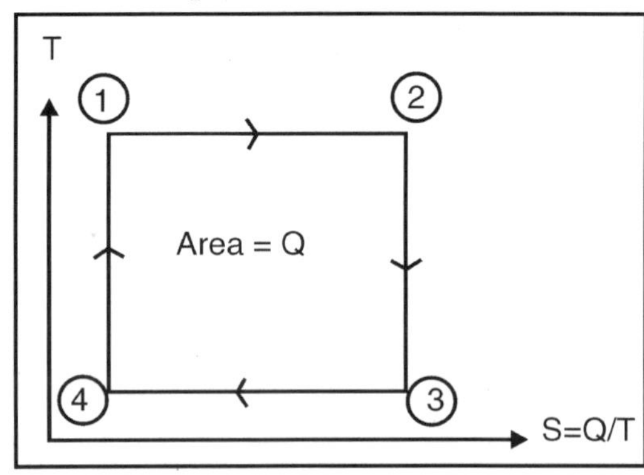

The change in entropy (δS) of a substance is that quantity which when multiplied by the absolute temperature at which the change took place results in the amount of energy (δQ) flow reversibly by heat transfer across the boundary enclosing the substance.

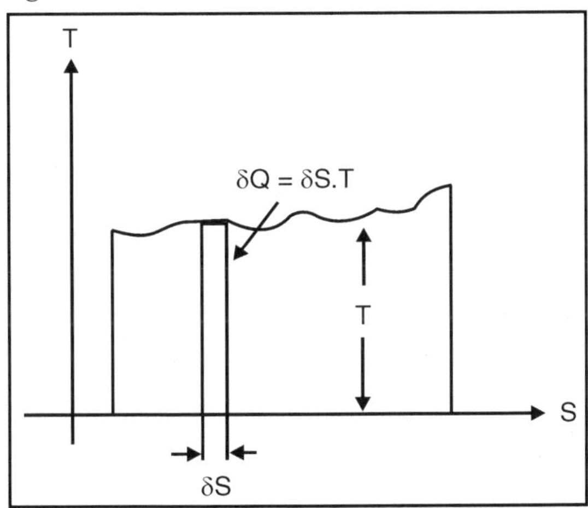

Total entropy is a property of a substance and therefore the change in entropy during a process, from an initial to a final state, is the same whatever the path taken. Therefore to determine the change in entropy resulting from a real irreversible process an equivalent reversible process must be envisaged to replace the real process from initial to final states before integration of the following

$$\int_i^f \frac{dQ_{rQv}}{T} = S_f - S_i$$

Typical SI units for total entropy (S) change are kJ /K. SI units for specific entropy (s) change are kJ/kg.K

Normally, students have a problem with the concept of entropy. This is expected, but what is unfortunate is that they think they understand energy. One supposes that if we use a term enough we think we understand it. After the population crisis and very much related to it, the most serious crisis facing humanity is usually referred to as the energy crisis – even by the President of the United States, high government officials, and famous professors.

Of course it should be referred to as the entropy crisis, availability crisis, or emergy crisis, but we are getting a little ahead of ourselves. (It came as quite a shock to one of us when he discovered that the people who are running the world don't know what they are talking about – much less what they are doing.) To facilitate the definition of entropy without recourse to the Second Law, we shall depart from so-called classical thermodynamics, which forbids inquiry into the microscopic nature of the universe. Just for a moment, we shall take a quick peek at statistical thermomechanics.

Typically, a system the entropy of which we wish to know has been defined in terms of common macroscopic thermodynamic properties that we have at our disposal and with which the reader may already be familiar such as volume, pressure, temperature, and internal energy for which we have *not* given formal definitions. In addition, let us suppose that the analyst is in possession (hypothetically) of a number of probability distributions $D_k = \{p_i, i = 1, 2, \ldots, N_k\}$ each of which, k = 1,..., M, (i) provides the probability that the system will be found in the *i*-th *microscopic* state, and (ii) is entirely consistent with the known*macroscopic* properties of the system.

Remember, in classical thermodynamics, we do not inquire deeply into the microscopic picture; therefore; we should not be surprised to find that more than one – perhaps many – such probability distributions could represent precisely the same state viewed macroscopically. For each such distribution a candidate S_k can be calculated for the entropy of the system. It is the minimal amount of information – measured in bits, say – to determine from the distribution under investigation the exact microscopic state of the system. It should not be construed that this determination could actually be carried out – even theoretically; but, it is easy to determine the expected (in the probabilistic sense) information deficit corresponding to the known macroscopic thermodynamic variables and the *j*-th probability distribution

$$D_j = \{p_i \log p_i | i = 1, N_j\} = \left\{-\sum_{i=1}^{N_j} p_i \log_2(p_i)\right\}_j$$

where N_j different possible and compatible microscopic states are associated with D_j. The macroscopic entropy of the system, S, is the maximum value of the minimal information deficits, i.e., S = maximum$\{S_j, j = 1 \text{ to } M\}$, where M probability distributions are compatible with the macroscopic state of the system as described by classical thermodynamics. As a somewhat challenging exercise, the reader may show, by considering the old TV game *Twenty Questions*, that the minimal information to determine the exact microscopic state of the *i*-th microstate is the log to the base two of p_i. This will give the entropy as the amount of information needed in bits, which would be converted into standard thermodynamic units by multiplying the same amount of information expressed as a natural logarithm

$$\log_2 p_i = \frac{\text{In}\, p_i}{\text{In}\, 2} = \text{In}\left(\sqrt[h2]{p_i}\right)$$

by Boltzmann's constant, $k = R_o/A$, where R_o is the universal gas constant and A is Avogadro's number. (Don't worry if you don't know what these numbers are.) The standard units of entropy are *energy over temperature* – essentially for traditional reasons. Despite the unfortuitous happenstance that *Joules per Kelvin* is not particularly suggestive of information, entropy is a measure of the quantity of information that would be needed to determine the microstate from the classical thermodynamic macrostate (except for a constant factor) although it is sometimes referred to as a measure of uncertainty, disorder, randomness, or chaos, which, from our view, are less satisfactory interpretations. The preceding remarks on entropy are derived from Dr. David Bowman's generous postings to a list server for physics teachers on the Internet.

Let us continue our mathematical analysis for a simple and commonly encountered probability distribution. Suppose a system be capable of taking [capital omega] different quantum states with the probability for the *i*-th state being p_i. Then, the entropy of that system is S=-k S p_i *ln* p_i, where S represents the summation from i = 1 to i = Ω, *ln* is the natural logarithm [*ln* p_i is the exponent to which the transcendental number *e* (equal to approximately 2.718282) must be raised to get the number p_i] and *k* is Boltzmann's constant. The case where the probability of each quantum state is the same, namely, when $p_i = 1/\Omega$, is exceptionally famous. In that case, we get S = k *ln* Ω. The expression S = k ln Ω appears on Boltzmann's tomb – unless we have been hoodwinked by the scientific historians, which is not entirely out of the question, although Truesdell provides a photograph.

We may now distinguish between heat and work, both of which entail the transfer of energy at the boundary and only at the boundary of a control volume; however, heat carries entropy along with it and affects the entropy balance accordingly. Work, on the other hand, carries no entropy, and, therefore, has no effect on the entropy balance. Because of this difference between heat and work an asymmetry arises; namely, work can be converted entirely to heat in every case including the case where no other change occurs in the universe, but heat can be converted completely to work only when the rest of the universe is changed in some additional fundamental way. If we restrict ourselves to cyclic processes, work can be converted completely to heat but not vice versa.

This is easy to visualize: Case 1: Consider the spring escapement in your Grandfather's watch. It transfers work to the gears, pointers, and whatever else accepts the work done by the spring mechanism, which we will take as "the system". The state of the material objects by means of which this energy is transferred as work is completely organized. The microstate is completely determined by the extent to which the spring has become unwound.

The probability of being in that state is one and the log of one is zero. The entropy is zero. Case 2: Imagine, for a moment, a hydraulic watch, say, driven by a pumped hydraulic fluid, which is permitted to flow across the control surface as we conceive it. This appears to be work too, but the turbulence and random motion of the pumped fluid as well as the friction losses in the pipe convert the electricity that drives the pump (work) into part work and part heat due to the turbulence and other forms of fluid friction. Since heat is crossing the control surface it will be accompanied by entropy since clearly the fluid will *not* be found in a state clearly defined by one parameter that takes its unique value with probability one. Case 3: Finally, if we had a steam clock, we could drive it by heating water in a boiler the surface of which facing the fire is our conceptual control surface. This is a case of heat and only heat crossing the control surface. Consider the entropy associated with this heat. Hint: Imagine how complicated fire is. The complications in the nature of the fire will complicate the conduction of heat to the boiler.]

Some analysts may quibble that heat doesn't cross the control surface. Rather, *thermal energy*crossing the control surface *is* heat; that is, crossing the control surface is part of the description of heat not separate from it. We do not feel the necessity for this kind of precision.

ENTROPY CHANGE FOR IDEALLY CLOSED SYSTEM GAS

For an ideal constant volume process... ($dQ = mc_v dT$)

$$dS = \frac{dQ}{T} = \frac{mc_v dT}{T}$$

Integrating between initial (2) and final state (2)

$$S_2 - S_1 = mc_v \log_\theta \frac{T}{T_1}$$

For an ideal constant pressure process... ($dQ = mc_p dT$)

$$dS = \frac{dQ}{T} = \frac{mc_p dT}{T}$$

Integrating between initial (2) and final state (2)

$$S_2 - S_1 = mc_p \log_e \frac{T_2}{T_1}$$

For an ideal adiabetic process...

$$dQ = 0$$

$$dS = \frac{dQ}{T} = 0$$

For an ideal isothermal process...

There is no change in temperature and therefore there is no change in internal energy U.

From the first law of thermodynamic $dQ = dU + dW$. If $dU = 0$ then $dQ = dW$.

$$dS = \frac{dW}{T} = \frac{pdV}{T}$$

$$\text{as } pV = mRT \text{ or } T = \frac{pV}{mR}$$

$$dS = \frac{pdVmR}{pV} = \frac{mRdV}{V}$$

Integrating between initial (2) and final state (2)

$$S_2 - S_1 = mR\log_e \frac{V_2}{V_1}$$

Entropy changes in terms of properties of a perfect gas

From the definition of entropy i.e δQ_{rev} = TδS and from the first law of thermodynamics is dQ = dE + dW which, for a reversible process can be written as δQ_{rev} =δE + PδV... it can be deduced that dE = TdS - PdV for a reversible process and if the substance is a gas of mass m then

$$dQ = TdS = mc_v dT + PdV........\text{equation X}$$

For a perfect gas PV = mRT and assuming C_v is constant.

a) Entropy change in terms of V and T

$$\text{by substituting } P = \frac{RmT}{V} \text{ in equation x}$$

$$\int_1^2 dS = mC_v \int_1^2 \frac{dT}{T} + Rm \int_1^2 \frac{dV}{V}$$

$$S_2 - S_1 = mC_v \log_e \left(\frac{T_2}{T_1}\right) + Rm \log_e \left(\frac{V_2}{V_1}\right)$$

b) Entropy change in terms of P and T

By differentiating PV = mRT

$$PdV = RmdT - VdP = RmdT - \frac{RmT}{P} dP$$

Substituting above in x and integrating to get

$$\int_1^2 dS = m(C_v + R) \int_1^2 \frac{dT}{T} - Rm \int_1^2 \frac{dP}{P} \quad \text{or}$$

$$S_2 - S_1 = mC_P \log_e \left(\frac{T_2}{T_1}\right) - Rm \log_e \left(\frac{P_2}{P_1}\right)$$

c) Entropy change in terms of P and V

$$T = \frac{PV}{Rm} \text{ and differentiating } PV = mRT$$

$$\text{Result in } dT = \frac{PdV + VdP}{Rm}$$

Substituting above in x and integrating to get

$$\int_1^2 dS = m(C_v + R) \int_1^2 \frac{dV}{V} - mC_v \int_1^2 \frac{dP}{P} \quad \text{or}$$

$$S_2 - S_1 = mC_P \log_e \left(\frac{V_2}{V_1}\right) - mC_v \log_e \left(\frac{P_2}{P_1}\right)$$

THE GENERIC BALANCE EQUATION

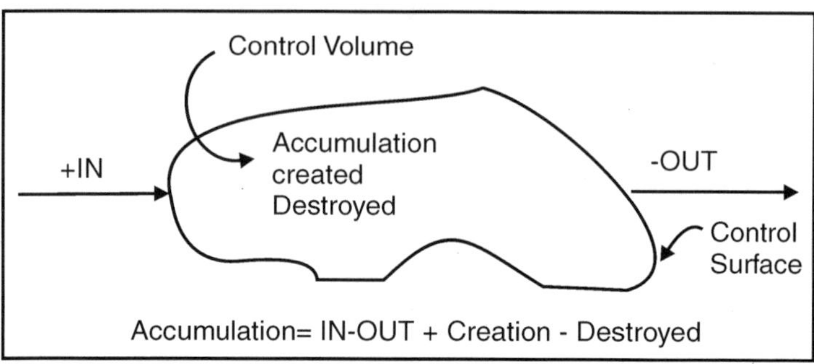

Fig. The Generic Balance Equation

The generic balance equation is so simple that, if we described it incorrectly to a three-year-old child, he or she would recognize that something was wrong, which is a good point in favor of the position that reasonableness is innate, i.e., an *a priori*synthetic judgment. One may argue as to what this equation may be applied to; but, if we should claim that it applies to what is commonly known as *stuff*, your objection would be a mere quibble. [Under the aegis of the commonality of the word stuff we consider both corporeal and incorporeal elements.

For example, energy is conserved, however it may be more like the behavior of a mysterious something than the something that behaves. It may be so abstract that nothing but a mathematical mapping of the relevant portion of the Universe is adequate to describe it. Mathematicians are content to refer to this mapping as *field equations*. As far as they are concerned, the field equations *are* the phenomenon.

The generic balance (or accounting) equation states quite simply that the accumulation within the control volume equals whatever is created within the control volume minus whatever is destroyed inside the control volume plus whatever enters the control volume minus whatever leaves the control volume. The situation is illustrated in Figure. By the accumulation we mean the difference between what we ended up with and what we started out with. This can be negative or positive; but, if it be negative, we might call its absolute value the deficit. As stated above, the control volume is any well-defined region in space. It may be changing shape and moving and it need not be a connected set.

Many purists will object that the balance equations are not the laws of thermodynamics, which, according to natural philosophy, must be statements that come entirely from experience and may not employ such abstract concepts as energy, temperature, and entropy. In particular, the Second Law should be a statement about a particular type of physical device that cannot exist. In fact, we have two statements each with its own impossible device. We suggest that the reader consult the excellent book for the layman by P.W. Atkins [3]. This will not be our last mention of this book. Suffice it to say that the balance-equation approach is logically, if not philosophically, equivalent to the experiential statements of the laws.]

We now wish to describe the balance equations of thermodynamics, which we are giving the status of laws, the famous laws of thermodynamics. We shall describe the First Law first. Since the First Law is an energy balance, since everyone thinks he knows what energy is, and since *entropy*, which is an important property of thermodynamic systems, does not appear in the First Law, most thermodynamics texts do not define entropy before they discuss the First Law of Thermodynamics, which involves both *work* and *heat*. Work and heat are *not* properties of the system, but they *are* rather subtle concepts.

Under some circumstances work and heat are mistaken for one another; whereas, if they are defined in terms of entropy, they can be distinguished easily – at least from the theoretical point of view. Therefore, we shall define and discuss entropy at this time. In our opinion, teachers of thermodynamics should give a little consideration to this departure from the usual way of presenting the First Law.

ENTROPY SYSTEM

Consider an isolated system whose energy is known to lie in a narrow range. Let Ω be the number of accessible microstates. According to the principle of equal a priori probabilities, the system is equally likely to be found in any one of these states when it is in thermal equilibrium. The accessible states are just that set of microstates which are consistent with the macroscopic constraints imposed on the system. These constraints can usually be quantified by specifying the values of some parameters $y_1, \ldots, y_n$ which characterize the macrostate. Note that these parameters are not necessarily external: e.g., we could specify either the volume (an external parameter) or the mean pressure (the mean force conjugate to the volume).

The number of accessible states is clearly a function of the chosen parameters, so we can write $\Omega \equiv \Omega\ (y_1, \ldots, y_n)$for the number of microstates consistent with a macrostate in which the general parameter y_α lies in the range y_α to $y_\alpha + dy_\alpha$.

Suppose that we start from a system in thermal equilibrium. According to statistical mechanics, each of the Ω_i, say, accessible states are equally likely. Let us now remove, or relax, some of the constraints imposed on the system. Clearly, all of the microstates formally accessible to the system are still accessible, but many additional states will, in general, become accessible. Thus, removing or relaxing constraints can only have the effect of increasing, or possibly leaving unchanged, the number of microstates accessible to the system. If the final number of accessible states is Ω_f, then we can write

$$\Omega_f \geq \Omega_i.$$

Immediately after the constraints are relaxed, the systems in the ensemble are not in any of the microstates from which they were previously excluded. So the systems only occupy a fraction

$$P_i = \frac{\Omega_i}{\Omega_f}$$

of the Ω_f states now accessible to them. This is clearly not a equilibrium situation. Indeed, if $\Omega_f \gg \Omega_i$ then the configuration in which the systems are only distributed over the original Ω_i states is an extremely unlikely one. In fact, its probability of occurrence is given by Eq. According to the theorem, *H* the ensemble will evolve in time until a more probable final state is reached in which the systems are evenly distributed over the Ω_f available states.

As a simple example, consider a system consisting of a box divided into two regions of equal volume. Suppose that, initially, one region is filled with gas and the other is empty. The constraint imposed on the system is, thus, that the coordinates of all of the molecules must lie within the filled region.

In other words, the volume accessible to the system $V = V_i$ is, where V_i is half the volume of the box. The constraints imposed on the system can be relaxed by removing the partition and allowing gas to flow into both regions. The volume accessible to the gas is now $V = V_f = 2V_i$. Immediately after the partition is removed, the system is in an extremely improbable state that at constant energy the variation of the number of accessible states of an ideal gas with the volume is

$$\Omega \propto V^N,$$

where N is the number of particles. Thus, the probability of observing the state immediately after the partition is removed in an ensemble of equilibrium systems with volume $V = V_f$ is

$$P_i = \frac{\Omega_i}{\Omega_f} = \left(\frac{V_i}{V_f}\right)^N = \left(\frac{1}{2}\right)^N.$$

If the box contains of order 1 mole of molecules then $N \sim 10^{24}$ and this probability is fantastically small:

$$P_i \sim \exp(-10^{24}).$$

Clearly, the system will evolve towards a more probable state.

This can also be phrased in terms of the parameters $y_1,...,y_n$ of the system. Suppose that a constraint is removed. For instance, one of the parameters, y, say, which originally had the value $y = y_i$, is now allowed to vary. According to statistical mechanics, all states accessible to the system are equally likely. So, the probability $P(y)$of finding the system in equilibrium with the parameter in the range to is just proportional y to $y + \delta y$ the number of microstates in this interval: i.e.,

$$P(y) \propto \Omega(y).$$

Usually, $\Omega(y)$ has a very pronounced maximum at some particular value $\tilde{y}$. This means that practically all systems in the final equilibrium ensemble have values of y close to $\tilde{y}$. Thus, if $y_i \neq \tilde{y}$ initially then the parameter y will change until it attains a final value close to $\tilde{y}$, where Ω is maximum.

If some of the constraints of an isolated system are removed then the parameters of the system tend to readjust themselves in such a way that

$$\Omega(y_1,...,y_n) \to \text{maximum}.$$

Suppose that the final equilibrium state has been reached, so that the systems in the ensemble are uniformly distributed over the Ω_f accessible final states. If the original constraints are reimposed then the systems in the ensemble still occupy these Ω_f states with equal probability. Thus, if $\Omega_f > \Omega_i$, simply restoring the constraints does not restore the initial situation. Once the systems are randomly distributed over the Ω_f states they cannot be expected to spontaneously move out of some of these states and occupy a more restricted class of states merely in response to the reimposition of a constraint. The initial condition can also not be restored by removing further constraints. This could only lead to even more states becoming accessible to the system.

Suppose that some process occurs in which an isolated system goes from some initial configuration to some final configuration. If the final configuration is such that the imposition or removal of constraints cannot by itself restore the initial condition then the process is deemed irreversible. On the other hand, if it is such that the imposition or removal of constraints can restore the initial condition then the process is deemed reversible.

From what we have already said, an irreversible process is clearly one in which the removal of constraints leads to a situation where $\Omega_f > \Omega_i$. A reversible process corresponds to the special case where the removal of constraints does not change the number of accessible states, so that $\Omega_f = \Omega_i$. In this situation, the systems remain distributed with equal probability over these states irrespective of whether the constraints are imposed or not.

Our microscopic definition of irreversibility is in accordance with the macroscopic. Recall that on a macroscopic level an irreversible process is one which "looks unphysical" when viewed in reverse. On a microscopic level it is clearly plausible that a system should spontaneously evolve from an improbable to a probable configuration in response to the relaxation of some constraint.

However, it is quite clearly implausible that a system should ever spontaneously evolve from a probable to an improbable configuration. Let us consider our example again. If a gas is initially restricted to one half of a box, via a partition, then the flow of gas from one side of the box to the other when the partition is removed is an irreversible process.

This process is irreversible on a microscopic level because the initial configuration cannot be recovered by simply replacing the partition. It is irreversible on a macroscopic level because it is obviously unphysical for the molecules of a gas to spontaneously distribute themselves in such a manner that they only occupy half of the available volume.

It is actually possible to quantify irreversibility. In other words, in addition to stating that a given process is irreversible, we can also give some indication of how irreversible it is. The parameter which measures irreversibility is just the number of accessible states Ω. Thus, if Ω for an isolated system spontaneously increases then the process is irreversible, the degree of irreversibility being proportional to the amount of the increase. If Ω stays the same then the process is reversible. Of course, it is unphysical for Ω to ever spontaneously decrease. In symbols, we can write

$$\Omega_f - \Omega_i \equiv \Delta\Omega \geq 0,$$

for any physical process operating on an isolated system. In practice, Ω itself is a rather unwieldy parameter with which to measure irreversibility. For instance, in the previous example, where an ideal gas doubles in volume (at constant energy) due to the removal of a partition, the fractional increase in Ω is

$$\frac{\Omega_f}{\Omega_i} \simeq 10^{2v\times 10^{23}},$$

where is the number of moles. This is an extremely large number! It is far more convenient to measure irreversibility in terms of ln Ω. If Eq. is true then it is certainly also true that

$$\ln \Omega_f - \ln \Omega_i \equiv \Delta \ln \Omega \geq 0$$

for any physical process operating on an isolated system. The increase in when an ideal gas doubles in volume (at constant energy) is

$$\ln \Omega_f - \ln \Omega_i = \nu\, N_A \ln 2,$$

where $N_A = 6 \times 10^{23}$. This is a far more manageable number! Since we usually deal with particles by the mole in laboratory physics, it makes sense to pre-multiply our measure of irreversibility by a number of order $1/N_A$. For historical reasons, the number which is generally used for this purpose is the Boltzmann constant k, which can be written

$$k = \frac{R}{N_A} \text{ joules/kelvin},$$

where

$$\text{R} = 8.3143 \text{ joules/kelvin/mole}$$

is the ideal gas constant which appears in the well-known equation of state for an ideal gas, $PV = v\,RT$. Thus, the final form for our measure of irreversibility is

$$S = k \ln \Omega$$

This quantity is termed "entropy", and is measured in joules per degree kelvin. The increase in entropy when an ideal gas doubles in volume (at constant energy) is

$$S_f - S_i = v\,R \ln 2,$$

which is order unity for laboratory scale systems (i.e., those containing about one mole of particles). The essential irreversibility of macroscopic phenomena can be summed up as follows:

$$S_f - S_i = \Delta S \geq 0,$$

for a process acting on an isolated system. Thus, the entropy of an isolated system tends to increase with time and can never decrease. This proposition is known as the second law of thermodynamics.

One way of thinking of the number of accessible states Ω is that it is a measure of the disorder associated with a macrostate. For a system exhibiting a high degree of order we would expect a strong correlation between the motions of the individual particles. For instance, in a fluid there might be a strong tendency for the particles to move in one particular direction, giving rise to an ordered flow of the system in that direction. On the other hand, for a system exhibiting a low degree of order we expect far less correlation between the motions of individual particles.

It follows that, all other things being equal, an ordered system is more constrained than a disordered system, since the former is excluded from microstates in which there is not a strong correlation between individual particle motions, whereas the latter is not. Another way of saying this is that an ordered system has less accessible microstates than a corresponding disordered system. Thus, entropy is effectively a measure of the disorder in a system (the disorder increases with S). With this interpretation, the second law of thermodynamics reduces to the statement that isolated systems tend to become more disordered with time, and can never become more ordered.

Note that the second law of thermodynamics only applies to isolated systems. The entropy of a non-isolated system can decrease. For instance, if a gas expands (at constant energy) to twice its initial volume after the removal of a partition, we can subsequently recompress the gas to its original volume. The energy of the gas will increase because of the work done on it during compression, but if we absorb some heat from the gas then we can restore it to its initial state. Clearly, in restoring the gas to its original state, we have restored its original entropy.

This appears to violate the second law of thermodynamics because the entropy should have increased in what is obviously an irreversible process (just try to make a gas spontaneously occupy half of its original volume!). However, if we consider a new system consisting of the gas plus the compression and heat absorption machinery, then it is still true that the entropy of this system (which is assumed to be isolated) must increase in time.

Thus, the entropy of the gas is only kept the same at the expense of increasing the entropy of the rest of the system, and the total entropy is increased. If we consider the system of everything in the Universe, which is certainly an isolated system since there is nothing outside it with which it could interact, then the second law of thermodynamics becomes:

The disorder of the Universe tends to increase with time and can never decrease. An irreversible process is clearly one which increases the disorder of the Universe, whereas a reversible process neither increases nor decreases disorder. This definition is in accordance with our previous definition of an irreversible process as one which "does not look right" when viewed backwards.

One easy way of viewing macroscopic events in reverse is to film them, and then play the film backwards through a projector. There is a famous passage in the novel "Slaughterhouse 5," by Kurt Vonnegut, in which the hero, Billy Pilgrim, views a propaganda film of an American World War II bombing raid on a German city in reverse. This is what the film appeared to show:

"American planes, full of holes and wounded men and corpses took off backwards from an airfield in England. Over France, a few German fighter planes flew at them backwards, sucked bullets and shell fragments from some of the planes and crewmen. They did the same for wrecked American bombers on the ground, and those planes flew up backwards and joined the formation.

The formation flew backwards over a German city that was in flames. The bombers opened their bomb bay doors, exerted a miraculous magnetism which shrunk the fires, gathered them into cylindrical steel containers, and lifted the containers into the bellies of the planes. The containers were stored neatly in racks. The Germans had miraculous devices of their own, which were long steel tubes. They used them to suck more fragments from the crewmen and planes. But there were still a few wounded Americans, though, and some of the bombers were in bad repair. Over France, though, German fighters came up again, made everything and everybody as good as new."

Vonnegut's point, I suppose, is that the morality of actions is inverted when you view them in reverse.

What is there about this passage which strikes us as surreal and fantastic? What is there that immediately tells us that the events shown in the film could never happen in reality? It is not so much that the planes appear to fly backwards and the bombs appear to fall upwards. After all, given a little ingenuity and a sufficiently good pilot, it is probably possible to fly a plane backwards.

Likewise, if we were to throw a bomb up in the air with just the right velocity we could, in principle, fix it so that the velocity of the bomb matched that of a passing bomber when their paths intersected. Certainly, if you had never seen a plane before it would not be obvious which way around it was supposed to fly. However, certain events are depicted in the film, "miraculous" events in Vonnegut's words, which would immediately strike us as the wrong way around even if we had never seen them before. For instance, the film might show thousands of white hot bits of shrapnel approach each other from all directions at great velocity, compressing an explosive gas in the process, which slows them down such that when they meet they fit together exactly to form a metal cylinder enclosing the gases and moving upwards at great velocity. What strikes us as completely implausible about this event is the spontaneous transition from the disordered motion of the gases and metal fragments to the ordered upward motion of the bomb.

Properties of Entropy

Entropy, as we have defined it, has some dependence on the resolution δE to which the energy of macrostates is measured. Recall that $\Omega(E)$ is the number of accessible microstates with energy in the range E to $E + \delta E$. Suppose that we choose a new resolution $\delta^* E$ and define a new density of states $\Omega^* E$ which is the number of states with energy in the range E to $E + \delta^* E$. It can easily be seen that

$$\Omega^*(E) = \frac{\delta^* E}{\delta E}\Omega(E).$$

It follows that the new entropy $S^* = k \ln \Omega^*$ is related to the previous entropy $S^* = k \ln \Omega$ via

$$S^* = S + k \ln \frac{\delta^* E}{\delta E}.$$

Now, our usual estimate that $\Omega \sim E^f$ gives $S \sim kf$, where f is the number of degrees of freedom. It follows that even if $\delta^* E$ were to differ from δE by of order f (i.e., twenty four orders of magnitude), which is virtually inconceivable, the second term on the right-hand side of the above equation is still only of order $k \ln f$, which is utterly negligible compared to kf. It follows that

$$S^* = S$$

to an excellent approximation, so our definition of entropy is completely insensitive to the resolution to which we measure energy (or any other macroscopic parameter).

Note that, like the temperature, the entropy of a macrostate is only well-defined if the macrostate is in equilibrium. The crucial point is that it only makes sense to talk about the number of accessible states if the systems in the ensemble are given sufficient time to thoroughly explore all of the possible microstates consistent with the known macroscopic constraints. In other words, we can only be sure that a given microstate is inaccessible when the systems in the ensemble have had ample opportunity to move into it, and yet have not done so. Note that for an equilibrium state, the entropy is just as well-defined as more familiar quantities such as the temperature and the mean pressure.

Consider, again, two systems A and A' which are in thermal contact but can do no work on one another. Let E and E' be the energies of the two systems, and $\Omega'(E)$ and $\Omega'(E')$ the respective densities of states. Furthermore, let $E^{(0)}$ be the conserved energy of the system as a whole and the $\Omega^{(0)}$ corresponding density of states. We have from Eq. that

$$\Omega^{(0)}(E) = \Omega(E)\,\Omega'(E'),$$

where $E' = E^{(0)} - E$. In other words, the number of states accessible to the whole system is the product of the numbers of states accessible to each subsystem, since every microstate of A can be combined with every microstate of $A'$to form a distinct microstate of the whole system that in equilibrium the mean energy of A takes the value $\bar{E} = \tilde{E}$ for which $\Omega^{(0)}(E)$ is maximum, and the temperatures of A and A' are equal. The distribution of E around the mean value is of order $\Delta^*E = \tilde{E}/\sqrt{f}$, where f is the number of degrees of freedom. It follows that the total number of accessible microstates is approximately the number of states which lie within Δ^*E of $\tilde{E}$. Thus,

$$\Omega_{tot}^{(0)} \simeq \frac{\Omega^{(0)}(\tilde{E})}{\delta E} \Delta^* E.$$

The entropy of the whole system is given by

$$S^{(0)} = k \ln \Omega_{tot}^{(0)} = k \ln \Omega^{(0)}(\tilde{E}) + k \ln \frac{\Delta^* E}{\delta E}.$$

According to our usual estimate, $\Omega \sim E^f$, the first term on the right-hand side is of order kf whereas the second term is of order $k \ln (\tilde{E}/\sqrt{f}\,\delta E)$. Any reasonable choice for the energy subdivision δE should be greater than $\tilde{E}/f$, otherwise there would be less than one microstate per subdivision. It follows that the second term is less than or of order $k \ln f$, which is utterly negligible compared to kf. Thus,

$S^{(0)} = k \ln \Omega^{(0)} (\tilde{E}) = k\ln [\Omega(\tilde{E})\Omega(\tilde{E}')] = k \ln \Omega(\tilde{E}) + k \ln \Omega'(\tilde{E}')$

to an excellent approximation, giving

$$S^{(0)} = S(\tilde{E}) + S'(\tilde{E}').$$

It can be seen that the probability distribution for $\Omega^{(0)}(E)$is so strongly peaked around its maximum value that, for the purpose of calculating the entropy, the total number of states is equal to the maximum number of states [i.e., $\Omega_{tot}^{(0)} \sim \Omega^{(0)}(\tilde{E})$]. One consequence of this is that the entropy has the simple additive property Eq. Thus, the total entropy of two thermally interacting systems in equilibrium is the sum of the entropies of each system in isolation.

Uses of Entropy

We have defined a new function called entropy, denoted S, which parameterizes the amount of disorder in a macroscopic system. The entropy of an equilibrium macrostate is related to the number of accessible microstates Ω via

$$S = k \ln \Omega.$$

On a macroscopic level, the increase in entropy due to a quasi-static change in which an infinitesimal amount of heat $đQ$ is absorbed by the system is given by

$$dS = \frac{đQ}{T},$$

where T is the absolute temperature of the system. The second law of thermodynamics states that the entropy of an isolated system can never spontaneously decrease. Let us now briefly examine some consequences of these results.

Consider two bodies, A and A', which are in thermal contact but can do no work on one another. We know what is supposed to happen here. Heat flows from the hotter to the colder of the two bodies until their temperatures are the same. Consider a quasi-static exchange of heat between the two bodies. According to the first law of thermodynamics, if an infinitesimal amount of heat $đQ$ is absorbed by A then infinitesimal heat $đQ' = -đQ$ is absorbed by A'. The increase in the entropy of system A is $dS = đQ/T$ and the corresponding increase in the entropy of A' is $dS = đQ/T'$. Here, T and T' are the temperatures of the two systems, respectively. Note that $đQ$ is assumed to the sufficiently small that the heat transfer does not substantially modify the temperatures of either system. The change in entropy of the whole system is

$$dS^{(0)} = dS + dS' = \left(\frac{1}{T} - \frac{1}{T'}\right) đQ,$$

This change must be positive or zero, according to the second law of thermodynamics, so $dS^{(0)} \geq 0$. It follows that $đQ$ is positive (i.e., heat flows from A' to A) when $T' > T$, and vice versa. The spontaneous flow of heat only ceases when $T = T'$. Thus, the direction of spontaneous heat flow is a consequence of the second law of thermodynamics.

Note that the spontaneous flow of heat between bodies at different temperatures is always an irreversible process which increases the entropy, or disorder, of the Universe.

Consider, now, the slightly more complicated situation in which the two systems can exchange heat and also do work on one another via a movable partition. Suppose that the total volume is invariant, so that

$$V^{(0)} = V + V' \text{ constant},$$

where V and V' are the volumes of A and A', respectively. Consider a quasi-static change in which system A absorbs an infinitesimal amount of heat $đQ$ and its volume simul-taneously increases by an infinitesimal amount dV. The infinitesimal amount of work done by system A is $đW = \bar{p}dV$, where $\bar{p}$ is the mean pressure of A. According to the first law of thermodynamics,

$$đQ = dE + đW = dE + \bar{p}dV,$$

where dE is the change in the internal energy of A. Since $dS = đQ/T$, the increase in entropy of system A is written

$$dS = \frac{dE + \bar{p}dV}{T}.$$

Likewise, the increase in entropy of system A' is given by

$$dS' = \frac{dE' + \overline{p}'dV'}{T'}.$$

According to Eq.

$$\frac{1}{T} = \left(\frac{\partial S}{\partial E}\right)_V,$$

$$\frac{\overline{p}}{T} = \left(\frac{\partial S}{\partial V}\right)_E,$$

where the subscripts are to remind us what is held constant in the partial derivatives. We can write a similar pair of equations for the system A'.

The overall system is assumed to be isolated, so conservation of energy gives $dE + dE' = 0$. Furthermore, Eq. implies that $dV + dV' = 0$. It follows that the total change in entropy is given by

$$dS^{(0)} = dS + dS' = \left(\frac{1}{T} - \frac{1}{T'}\right)dE + \left(\frac{\overline{p}}{T} - \frac{\overline{p}'}{T'}\right)dV.$$

The equilibrium state is the most probable state. According to statistical mechanics, this is equivalent to the state with the largest number of accessible microstates. Finally, Eq. implies that this is the maximum entropy state.

The system can never spontaneously leave a maximum entropy state, since this would imply a spontaneous reduction in entropy, which is forbidden by the second law of thermodynamics.

A maximum or minimum entropy state must satisfy $dS^{(0)} = 0$ for arbitrary small variations of the energy and external parameters. It follows from Eq. that

$$T = T'$$

$$\overline{p} = \overline{p}',$$

for such a state. This corresponds to a maximum entropy state (i.e., an equilibrium state) provided

$$\left(\frac{\partial^2 S}{\partial E^2}\right)_V < 0,$$

$$\left(\frac{\partial^2 S}{\partial V^2}\right)_E < 0,$$

with a similar pair of inequalities for system A'. The usual estimate $\Omega \propto E^f V^f$, giving $S = kf \ln E + kf \ln V + \ldots$, ensures that the above inequalities are satisfied in conventional macroscopic systems. In the maximum entropy state the systems A and A' have equal temperatures (i.e., they are in thermal equilibrium) and equal pressures (i.e., they are in mechanical equilibrium). The second law of thermodynamics implies that the two interacting systems will evolve towards this state, and will then remain in it indefinitely (if left undisturbed).

ENTROPY MESSAGE

Take-home message: The second law leads us to define a new function of state, entropy. The entropy of an isolated system can never decrease.

Clausius's theorem says that if a system is taken through a cycle, the sum of the heat added weighted by the inverse of the temperature at which it is added is less than or equal to zero:

$$\oint \frac{đQ}{T} \leq 0$$

This follows from Clausius's statement of the second law. The details of the proof are here. The inequality becomes an equality for reversible systems:

$$\oint \frac{đQ^{rev}}{T} = 0$$

PROOF OF CLAUSIUS'S THEOREM

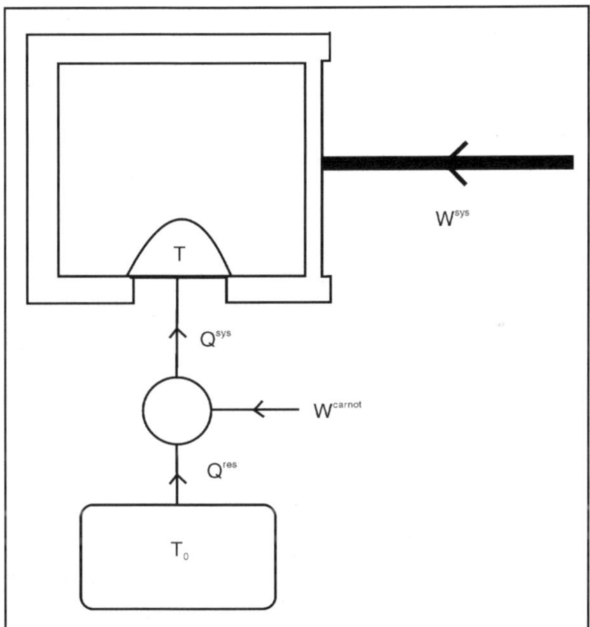

In the diagram, the system is the gas in the piston. We use a Carnot heat engine/pump to add heat Q^{sys} to the system at a local, varying temperature T. During the process work W^{says} is done on the system. (Both Q^{sys} and W^{says} could be positive or negative.) The cold reservoir of the Carnot engine is at T_0.

The Kelvin-Planck statement of the second law says that at the end of a complete cycle of the system, we cannot have extracted net work from the system (or else we would have turned heat into work). Looking at the figure to see how the signs of the various works are defined, that means $W^{carnot}+W^{sys}>0$. By conservation of energy, and because the system and engine have returned to their initial states, and net work put in must end up as heat added to the reservoir: $Q^{res}\leq 0$ (less than zero because Q^{res} is defined as heat extracted).

Looking now at the Carnot engine, we see that if we add heat $đQ^{sys}$ at temperature T, heat $đQ^{res}$ is extracted from the reservoir, and

$$đQ^{res} = \frac{T_0}{T} đQ^{sys}$$

Since the total heat extracted is less than zero, we have

$$Q^{res} = \oint đQ^{res} = \oint \frac{T_0}{T} đQ^{sys} = T_0 \oint \frac{đQ^{sys}}{T} \leq 0$$

which proves the inequality: (dropping the superscript "sys")

$$\oint \frac{đQ}{T} \leq 0$$

Clearly if the system is taken through a reversible cycle, it can be run in reverse and all quantities will simply change signs. But if Q^{res} was less than zero originally, it will be greater for the reversed cycle, implying a net extraction of work and violating the Kelvin-Planck statement. Thus for a reversible system, Q^{res} must be exactly zero, and

$$\oint \frac{đQ}{T} \doteq 0$$

We can verify that this holds for a system which is taken through a Carnot cycle, since there heat only enters or leave at one of two temperatures:

$$\oint \frac{đQ^{rev}}{T} = \frac{Q_H}{T_H} - \frac{Q_C}{T_C} = 0$$

(A less rigorous proof of Clausius's theorem, used in Bowley and Sánchez and in Zemansky, involves approximating any reversible cycle by a large number of Carnot cycles.)

This is interesting because a quantity whose change vanishes over a cycle implies a function of state. We know that heat itself isn't a function of state, but it seems that in a reversible process "heat over temperature" is a function of state. It is called entropy with the symbol S:

$$dS = \frac{đQ^{rev}}{T}$$

and .

$$\oint dS = 0$$

So much for cycles. What about other processes? By considering a cycle consisting of one reversible and one irreversible process, we can show that in general,

$$đQ \leq TdS$$

This gives rise to the most important result of all. For an isolated system, $đQ = 0$. So for such a system,

$$dS \geq 0$$

This leads to an alternative statement of the second law: The entropy of an isolated system can never decrease.

A system and its surroundings together ("the universe") form an isolated system, whose entropy never decreases: any decrease in the entropy of a system must be compensated by the entropy increase of its surroundings.

Increase of Entropy

Here we imagine a cycle in which we go from point 1 to point 2 by an irreversible process and back by a reversible process.

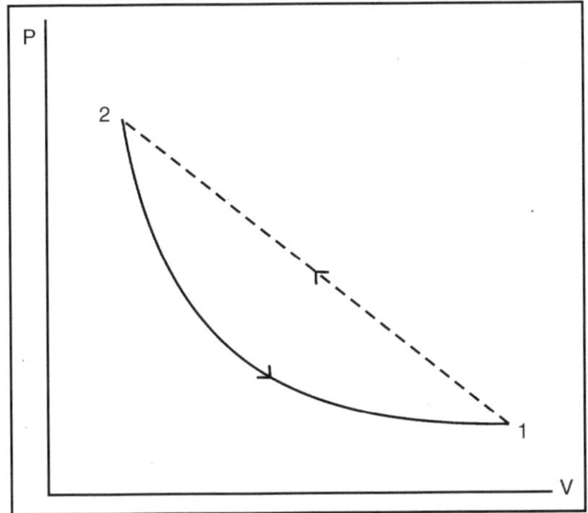

The entropy change for a reversible process is given by

$$S_2 - S_1 = \int_1^2 \frac{đQ^{rev}}{T}$$

What can we say about an irreversible process between the same endpoints? Clearly the entropy change is the same (that's what we mean by saying entropy is a function of state.) But if we consider a cycle involving an irreversible process from 1 to 2 and a reversible process to return to 1, we have a cycle for which Clausius's theorem holds:

$$\oint \frac{đQ}{T} < 0$$

$$\Rightarrow \int_1^2 \frac{đQ^{irrev}}{T} + \int_2^1 \frac{đQ^{rev}}{T} < 0$$

$$\Rightarrow \int_1^2 \frac{đQ^{irrev}}{T} < S_2 - S_1$$

Hence in general,

$$đQ \leq TdS$$

and so for an isolated system, where $đQ = 0$,

$$dS \geq 0$$

THE ENTROPY BALANCE AND LOST WORK

We are now ready to write the Second Law of Thermodynamics as an entropy balance:

Entrophy Balance:

$$\Delta S = m_2 s_2 - m_1 s_1$$

$$= \left(\sum m_i s_i + \sum \frac{Q_i}{T_i}\right) - \left(\sum m_e s_e + \sum \frac{Q_e}{T_e}\right) + \frac{L_{CV}}{T_0}$$

As before, associated with each heat term Q_i accounting for energy and entropy entering the control volume is a temperature T_i normally the temperature at the control surface. Similarly, the subscript *e* refers to heat accounting for energy and entropy leaving the control volume (CV). If the temperature be not constant, we employ the integral calculus. The symbol T_o represents the temperature of the surroundings of the CV – assumed to be constant.

Normally, this is the temperature of the air or a convenient body of water. In most engineering calculations, we will not make a significant error if we take T_O to be 288 Kelvin (written 288 K – *without* a degree sign) everywhere on the Earth both summer and winter. (Temperature in Kelvin is degrees Celsius (Centigrade) plus about 273.16.) Eventually, we shall be comparing the temperature of the Earth to the temperature of the Sun and what might seem to be extreme differences in temperatures if one had to subject one's body to them will be insignificant mathematically.

Therefore, we shall assume that the temperature of the Earth is constant at 288 K. However, when the CV is the entire Earth and a shell surrounding it 100 miles thick, the temperature of the surroundings must be considered carefully. The expression L stands for *(thermodynamic) lost work,* which is really a very suggestive term. We shall see exactly what it represents in the next section on the First and Second Laws Combined. For now, it is what makes the Second Law an equation rather than an inequality. For that reason alone, it is an extremely important concept.

Just as in the case of the First Law the expression $m_2 s_2 - m_1 s_1$ represents the accumulation term. The two terms with summation signs represent entropy crossing the control surface (in and out, respectively). The terms $m_i s_i$ and $m_e s_e$ represent mass crossing the control surface each unit of which carries its own specific entropy, whereas the heat terms (ratios of heat to temperature) represent entropy crossing the boundary that is *not* associated with mass.

Notice, as mentioned earlier, the entropy balance has no work terms. (Why?) In this balance equation (unlike the First Law) we have a creation term, namely, L_{CV}/T_O. Since L_{cv} is always positive and T_o is always positive, this term always represents creation of entropy. It is this term L_{CV}/T_O that represents irreversibilities, I, in the process, i.e., $I = L_{CV}/T_O$. Examples of irreversibilities are friction, turbulence (in fluids), the mixing of pure substances, the unrestrained expansion of a gas, and transfer of heat over a finite temperature difference.

Thus, the heat terms, Q_{CS}/T_{CS}, in this version of the Second Law must represent *reversible* heat transfer, i.e., thermal energy that is exchanged infinitely slowly with a *thermal reservoir*. Considerations of reversibility (approachable but not obtainable by real processes), and irreversibility are of paramount importance in classical thermodynamics – as we shall see.

A thermal reservoir is a large thermal energy sink or thermal energy source that (i) is capable of exchanging essentially infinite thermal energy without changing temperature, i.e., is very large – like the entire atmosphere, and that (ii) differs in temperature only infinitesimally from the temperature at the control surface, T_{CS}, in the denominator of these terms. (We have employed the subscript *cs* when the direction of transfer isn't important.) That is, thermal energy can be exchanged reversibly from a thermal reservoir at $T_i + dT_i$ to a control surface at T_i. Likewise, from a control surface at T_e to a thermal reservoir at $T_e - dT_e$.

The classical example of irreversibility given in popular expositions is a glass falling from a table to the floor and breaking into a thousand pieces. (This is like *mechanical* lost work, where T is the temperature of the system and I is the irreversibility produced in the control volume. If we subtract the work required to clean up the mess of the broken glass (and the spilled wine, perhaps) from the mechanical lost work, *that* is analogous to the *thermodynamic* lost work that we are using in our version of the famous equation; i.e., $L_{thermo} = L_{CV} = T_o'I$.)

We do not expect to see this process reverse itself spontaneously unless someone is running a motion picture backwards. In fact that's how we know the motion picture is running backwards and it makes us laugh (or smile). This irreversibility of nearly all real processes is referred to as "the arrow of time" and we all believe in it (or we wouldn't smile at the motion picture running backward). Thus, at least in this part of the universe and during this era in the development of the universe, the Second Law tells us that the entropy of the universe is always increasing. This does not mean, of course, that the entropy of every control volume is increasing; but, if it's decreasing in one control volume, it's increasing even faster somewhere else. We shall consider the important concept of a Carnot engine next.

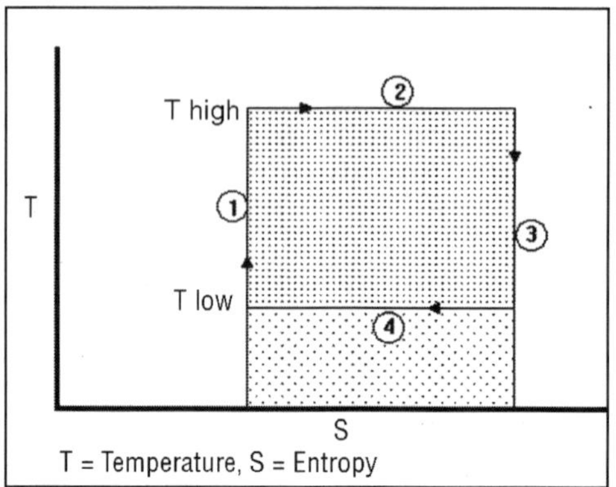

Fig. The Temperature-Entropy (S vs T) Diagram for a Carnot Engine

The thermodynamic cycle for an imaginary Carnot engine is pictured on an entropy-temperature diagram in Fig. The numbers in circles refer to the following process steps: (1) an isentropic (constant entropy) pumping of the imaginary working fluid (presumably a liquid) from a low-pressure, low-temperature state to a high-temperature, high-pressure state, (2) reversible heat exchange over an infinitesimal temperature difference from a high-temperature heat reservoir to the working fluid, which stays at constant temperature (presumably while the working fluid is changing from a liquid to a vapor), (3) an isentropic expansion of the fluid (presumably through a gas turbine, which delivers work, some of which is used in Step 1) from a high-temperature, high-pressure state to a low-pressure, low-temperature state, and (4) reversible heat exchange over an infinitesimal temperature difference at a constant low temperature (presumably while the working fluid is condensing from a vapor to a liquid).

The entire area within the shaded rectangles represents the heat exchanged at the high temperature; the lightly shaded rectangle represents the heat exchanged with the surroundings at the low temperature; the heavily shaded rectangle represents the work done by the imaginary heat engine, i.e., the difference between the heat in and the heat out.

Such a heat engine, operating in such a cycle, is called a Carnot engine, after Nicolas Leonard Sadi Carnot, a French physicist who was born in 1796 and died (young) in 1832. Clearly, a heat exchanger that exchanges heat through an infinitesimal temperature difference would have to have an infinite area, which is inconvenient for purposes of construction. Also, it is difficult to imagine what sort of fluid could go from low temperature to high temperature while being pumped as a liquid (this would be necessary to minimize the portion of work that would have to be drawn from the turbine to operate the device that brings the fluid from low pressure to high pressure). Nevertheless, the Carnot engine is a useful concept that represents an upper bound on efficiency for real heat engines. If someone tries to sell you a heat engine for which an efficiency better than the efficiency of a Carnot engine is claimed, "stay not on the order of your leaving, but depart immediately." – William Burroughs in *Naked Lunch*. In the book by P.W. Atkins [3] a much more credible Carnot cycle is illustrated. Any reversible cyclic engine can be a Carnot engine provided only that it has two isentropic processes and two isothermal processes. Atkins illustrates his Carnot engine with a pressure-volume diagram – the well-known indicator diagram employed by James Watt.

The formula for the work from a Carnot engine is easily derived from the Second Law for a *cyclic* reversible process with no material entering or leaving the system. Remember that, during one cycle of a cyclic process, the system is returned to the state from which it started. Therefore, the accumulation term must be zero. Also, for a reversible process, the lost work term is zero. Thus,

$$0 = \frac{Q_i}{T_i} - \frac{Q_e}{T_e}$$

To analyze this process denote the change in entropy of the system during Step 2 as the positive quantity ΔS. This is precisely equal to the positive change in entropy of the surroundings during Step 4, as is clear from Fig. The system gains entropy during the heat input step and loses the same amount of entropy during the heat rejection step, which results in no change over the course of one cycle consisting of all four steps. The heat associated with energy added at the high temperature (H is for high) is , while the heat associated with energy rejected at the low temperature (L is for low) is . The work done by the engine, then, is W_{rev} = , while the efficiency, η

Frequently, $T_L = T_o$ and T_H is just plain T, so

$$\eta = 1 - \frac{T_0}{T} .$$

The efficiency of a Carnot engine can be approached as closely as we are willing to pay for, but it can never be attained by a real engine.

EXAMPLES OF ENTROPY CHANGES

Take-home message: Spontaneous changes are aways associated with entropy increase. Below we have various examples of entropy change during various processes. From them we can draw some general conclusions. First we see in general the entropy of anything increases when it is heated, and the entropy of a gas increases when it expands at constant temperature. A common theme (to be explored in more detail later) is that an increase in entropy is associated with an increase in disorder.

Examples of spontaneous processes are the flow of heat from a hotter to a colder body (see Ex. 1) and the free expansion of a gas. In both of these, the total entropy increases (though that of parts of the system may decrease). This is completely general: spontaneous processes are those which increase entropy.

Since heat flow from hotter to colder bodies is irreversible, reversible processes must involve heat flow between bodies of the *same* temperature. It follows that any entropy change of the system $dS_{sys} = đQ^{rev}/T$ must be exactly balanced by that of the heat bath which provided the heat: $dS_{res} = -đQ^{rev}/T$. Thus the entropy change of the universe during reversible processes is zero.

During an adiabatic process no heat flows. Thus from $dS = đQ^{rev}/T$ we see that the entropy change of a system during for a reversible, adiabatic process is zero. But note that both qualifiers are needed; the entropy of a non-isolated system can change during a reversible process (and the entropy change of the surroundings will compensate), and an irreversible change to an isolated system will increase the entropy.

- The entropy of the universe always increases during spontaneous c
- During reversible changes the entropy of the system may change, but that of the universe stays constant.
- It follows that spontaneous changes are always irreversible
- During reversible, adiabatic changes the entropy of the system is a

Try These Problems

Ex. 1

Two identical blocks of iron, one at 100°C and the other at 0C, are brought into thermal contact. What happens? What is the total entropy change? (Assume the heat capacity of each block, , is constant over this temperature range, and neglect volume changes)

Answer:

Both blocks end up at 50°C and the entropy change is . More details here.

Two identical blocks of iron, one at 100°C and the other at 0°C, are brought into thermal contact. What happens? What is the total entropy change? (Assume the heat capacity of each block, , is constant over this temperature range, and neglect volume changes)

We know what happens: heat flows from the hot to the cold body till they reach the same temperature; conservation of energy requires that this will be at 50°C. Why does heat transfer occur? If heat $đQ$ is transferred from a hot body at T_H to a cold one at T_C, the entropy decrease of the hot body is $dS_H > đQ/T_H$, and the entropy increase of the cold body is $dS_C > đQ/T_C$. So the total entropy change is

$$dS = dS_H + dS_C > \left(\frac{1}{T_C} - \frac{1}{T_H}\right) đQ > 0$$

So the decrease in entropy of the hot block is more than compensated by the increase in entropy of the cold block. The spontaneous flow of heat is associated with an overall entropy increase, and the two blocks exchange heat till their combined entropy is maximised.

What is the overall entropy change for the total process? Here we have to use a trick. We can't calculate entropy changes for irreversible processes directly; we need to imagine a reversible process between the same endpoints. For the heating or cooling of a block, this would involve bringing it in contact with a series of heat baths at infinitesimally increasing or decreasing temperatures, so that the temperature difference between the heat bath and the block is always negligible and the entropy change is zero.

During this process, the heat transfer and the infinitesimal temperature change are related by $đQ^{rev} = CdT$ and so

$$\Delta S = \int_1^2 \frac{đQ^{rev}}{T} = C\int_{T_1}^{T_2} \frac{dT}{T} = C \ln\left(\frac{T_2}{T_1}\right)$$

Thus the total entropy change is

$$\Delta S = \Delta S_C + \Delta S_H = C \ln\left(\frac{T_f}{T_C}\right) + C \ln\left(\frac{T_f^2}{T_H T_C}\right)$$

$$= C \ln\left(\frac{323^2}{273 \times 373}\right)$$

$$= 0.024C$$

Ex. 2

Two identical blocks of iron, one at 100°C and the other at 0°C, are brought into thermal contact. What is the maximum work that can be extracted from the hot block in the absence of other heat sinks?

Answer:

The final temperature is 46°C and 14% of the energy lost by the hot block is available to do work. More details here.

Two identical blocks of iron, one at 100°C and the other at 0°C, are brought into thermal contact. what is the maximum work that can be extracted from the hot block in the absence of other heat sinks?

Remember, we can't just extract heat from the hot block and turn it into work; the entropy of the block would decrease without any compensating increase elsewhere. We need to add at least enough heat to the cold block so that its entropy increases by as much as that of the hot block decreases. (Work can always be used to do things which don't increase the entropy of the universe, such as lifting a weight.) Once the two blocks are at the same temperature, no further work can be extracted.

At that point the entropy change of the two blocks together, from the previous example, is

$$\Delta S = c \ln\left(\frac{T_f^2}{T_H T_C}\right) \geq 0$$

The lowest final temperature is that for which $\Delta S = 0$, *ie* $T_f = \sqrt{T_H T_C} = 319$ K =46°C. (Any lower and ΔS would go negative, which isn't allowed.) So although the total heat loss of the hot block is $Q_H=c(T_H-T_f)$, the work extracted is only the difference between this and the heat gained by the cold block, $Q_C=c(T_f-T_C)$, namely $W=c(T_H+T_C-2T_f)$ In this case, the efficiency is

$$\eta=\frac{W}{Q_H}=\frac{(T_H+T_C-2T_f)}{(T_H-T_f)}=0.144$$

Only 14% of the heat lost by the hot block was available to do work! Note that the maximum efficiency was obtained from a reversible process.

Ex. 3

Heat engines revisited From the law of non-decrease of entropy, show that the maximum efficiency of a heat engine operating between two reservoirs at T_H and T_C occurs when the engine is reversible.

Answer:

Heat engines revisited: From the law of non-decrease of entropy, show that the maximum efficiency of a heat engine operating between two reservoirs at T_H and T_C occurs when the engine is reversible.

This is a bit circular, as we used the properties of Carnot engines to derive this form of the second law! However we will see later in the course that it can be independently derived using statistical methods.

The change in entropy of the two reservoirs, which must be non-negative, is

$$\Delta S=\Delta S_C+\Delta S_H=\frac{Q_C}{T_C}-\frac{Q_H}{T_H}\geq 0 \Rightarrow \frac{Q_C}{Q_H}\geq\frac{T_C}{T_H}$$

with the equality being for $\Delta S=0$, *ie* for a reversible process. So the efficiency is

$$\eta=\frac{W}{Q_H}=1-\frac{Q_C}{Q_H}\leq 1-\frac{T_C}{T_H}$$

This is maximum when the equality is satisfied, *ie* for a reversible engine.

Ex.

4n moles of an ideal gas at temperature T_0 are originally confined to half of an insulated container by a partition. The partition is removed without doing any work. What is the change in entropy?

Answer:

$\Delta S=nR$ In 2. More details here.

More details here.

moles of an ideal gas at temperature are originally confined to half of an insulated container by a partition. The partition is removed without doing any work. What is the final change in entropy?

First, we must resist the temptation to say (because process is adiabatic) hence . In fact and this is not a reversible process. To solve this problem, we need to find a reversible process linking the same two endpoints. We've already seen this; the process is a reversible isothermal expansion For such an expansion so

This is rather subtle; make sure you understand the difference between the actual process (with and) and the reversible process for which we could calculate the entropy change (with). Remember, and are not functions of state - only the sum is.

Ex. 5

An insulated container is originally divided in half by a partition, and each half is originally occupied by moles of an different ideal gas at temperature . The partition is removed without doing any work. What is the change in entropy?

Answer:

An insulated container is originally divided in half by a partition, and each half is originally occupied by moles of an different ideal gas at temperature . The partition is removed without doing any work. What is the change in entropy?

Since these are ideal gases, they do not interact. Each species is oblivious of the existence of the other and the entropy change for each is just the same as in the free expansion of . Thus the total entropy change is given by . Note that the total mixing - which we know will happen eventually - is exactly the change which maximises the entropy.

ENTROPY IN QUANTUM MECHANICS

In quantum statistical mechanics, the concept of entropy was developed by John von Neumann and is generally referred to as "von Neumann entropy". Von Neumann established a rigorous mathematical framework for quantum mechanics with his work Mathematische Grundlagen der Quantenmechanik. He provided in this work a theory of measurement, where the usual notion of wave collapse is described as an irreversible process (the so called von Neumann or projective measurement). Using this concept, in conjunction with the density matrix he extended the classical concept of entropy into the quantum domain. It is well known that a Shannon based definition of information entropy leads in the classical case to the Boltzmann entropy.

It is tempting to regard the Von Neumann entropy as the corresponding quantum mechanical definition. But the latter is problematic from quantum information point of view. Consequently Stotland, Pomeransky, Bachmat and Cohen have introduced a new definition of entropy that reflects the inherent uncertainty of quantum mechanical states. This definition allows to distinguish between the minimum uncertainty entropy of pure states, and the excess statistical entropy of mixtures.

ENTROPY IN ASTROPHYSICS

In astrophysics, what is referred to as "entropy" is actually the adiabatic constant derived as follows. Using the first law of thermodynamics for a quasi-static, infinitesimal process for a hydrostatic system

$$dQ = dU - dW.$$

For an ideal gas in this special case, the internal energy, U, is only a function of T; therefore the partial derivative of heat capacity with respect to T is identically the same as the full derivative, yielding through some manipulation

$$dQ = C_V dT + PdV.$$

Further manipulation using the differential version of the ideal gas law, the previous equation, and assuming constant pressure, one finds

$$dQ = C_P dT - VdP.$$

For an adiabatic process dQ = 0 and recalling $\gamma = \frac{C_P}{C_V}$, one finds

$$VdP = C_P dT$$
$$PdV = - C_V dT$$
$$\frac{dP}{P} = -\frac{dV}{P}\gamma.$$

One can solve this simple differential equation to find

$$PV^{\gamma} = \text{constant} = K$$

This equation is known as an expression for the adiabatic constant, K, also called the adiabat. From the ideal gas equation one also knows

$$P = \frac{\rho k_B T}{\mu m_H},$$

where k_B is Boltzmann's constant. Substituting this into the above equation along with $V = [grams]/\rho$ and $\gamma = 5/3$ for an ideal monoatomic gas one finds

$$K = \frac{k_B T}{\mu m_H r^{2/3}},$$

where μ is the mean molecular weight of the gas or plasma; and μ_H is the mass of the Hydrogen atom, which is extremely close to the mass of the proton, m_p, the quantity more often used in astrophysical theory of galaxy clusters. This is what astrophysicists refer to as "entropy" and has units of [keV cm2]. This quantity relates to the thermodynamic entropy as

$$S = k_B \ln \Omega + S_0$$

where Ω, the density of states in statistical theory, takes on the value of K as defined above.

Energy Dispersal

The concept of entropy can be described qualitatively as a measure of energy dispersal at a specific temperature. Similar terms have been in use from early in the history of classical thermodynamics, and with the development of statistical thermodynamics and quantum theory, entropy changes have been described in terms of the mixing or "spreading" of the total energy of each constituent of a system over its particular quantized energy levels. Ambiguities in the terms disorder and chaos, which usually have meanings directly opposed to equilibrium, contribute to widespread confusion and hamper comprehension of entropy for most students. As the second law of thermodynamics shows, in an isolated system internal portions at different temperatures will tend to adjust to a single uniform temperature and thus produce equilibrium.

A recently developed educational approach avoids ambiguous terms and describes such spreading out of energy as dispersal, which leads to loss of the differentials required for work even though the total energy remains constant in accordance with the first law of thermodynamics. Physical chemist Peter Atkins, for example, who previously wrote of dispersal leading to a disordered state, now writes that "spontaneous changes are always accompanied by a dispersal of energy", and has discarded 'disorder' as a description.

ENTROPY AND INFORMATION THEORY

In information theory, entropy is the measure of the amount of information that is missing before reception and is sometimes referred to as Shannon entropy. Shannon entropy is a broad and general concept which finds applications in information theory as well as thermodynamics. It was originally devised by Claude Shannon in 1948 to study the amount of information in a transmitted message. The definition of the information entropy is, however, quite general, and is expressed in terms of a discrete set of probabilities pi. In the case of transmitted messages, these probabilities were the probabilities that a particular message was actually transmitted, and the entropy of the message system was a measure of how much information was in the message. For the case of equal probabilities (i.e. each message is equally probable), the Shannon entropy (in bits) is just the number of yes/no questions needed to determine the content of the message.

The question of the link between information entropy and thermodynamic entropy is a hotly debated topic. Some authors argue that there is a link between the two, while others will argue that they have absolutely nothing to do with each other.

The expressions for the two entropies are very similar. The information entropy H for equal probabilities pi = p is:

$$H = K \ln (1/p)$$

where K is a constant which determines the units of entropy. For example, if the units are bits, then K=1/ln. The thermodynamic entropy S , from a statistical mechanical point of view was first expressed by Boltzmann:

$$S = 1 \ln (1/p)$$

where p is the probability of a system being in a particular microstate, given that it is in a particular macrostate, and k is Boltzmann's constant. It can be seen that one may think of the thermodynamic entropy as Boltzmann's constant, divided by ln, times the number of yes/no questions that must be asked in order to determine the microstate of the system, given that we know the macrostate.

The link between thermodynamic and information entropy was developed in a series of papers by Edwin Jaynes beginning in 1957.

The problem with linking thermodynamic entropy to information entropy is that in information entropy the entire body of thermodynamics which deals with the physical nature of entropy is missing.

The second law of thermodynamics which governs the behavior of thermodynamic systems in equilibrium, and the first law which expresses heat energy as the product of temperature and entropy are physical concepts rather than informational concepts.

If thermodynamic entropy is seen as including all of the physical dynamics of entropy as well as the equilibrium statistical aspects, then information entropy gives only part of the description of thermodynamic entropy. Some authors, like Tom Schneider, argue for dropping the word entropy for the H function of information theory and using Shannon's other term "uncertainty" instead.

Ice Melting Example

The illustration for this article is a classic example in which entropy increases in a small 'universe', a thermodynamic system consisting of the 'surroundings' (the warm room) and 'system' (glass, ice, cold water). In this universe, some heat energy δQ from the warmer room surroundings (at 298 K or 25 °C) will spread out to the cooler system of ice and water at its constant temperature T of 273 K (0°C), the melting temperature of ice.

The entropy of the system will change by the amount dS = δQ/T, in this example δQ/273 K. (The heat δQ for this process is the energy required to change water from the solid state to the liquid state, and is called the enthalpy of fusion, i.e. the ΔH for ice fusion.) The entropy of the surroundings will change by an amount dS = "δQ/298 K. So in this example, the entropy of the system increases, whereas the entropy of the surroundings decreases.

It is important to realize that the decrease in the entropy of the surrounding room is less than the increase in the entropy of the ice and water: the room temperature of 298 K is larger than 273 K and therefore the ratio, (entropy change), of δQ/298 K for the surroundings is smaller than the ratio (entropy change), of δQ/273 K for the ice+water system.

To find the entropy change of our "universe", we add up the entropy changes for its constituents: the surrounding room and the ice+water. The total entropy change is positive; this is always true in spontaneous events in a thermodynamic system and it shows the predictive importance of entropy: the final net entropy after such an event is always greater than was the initial entropy.

As the temperature of the cool water rises to that of the room and the room further cools imperceptibly, the sum of the $\delta Q/T$ over the continuous range, at many increments, in the initially cool to finally warm water can be found by calculus. The entire miniature "universe", i.e. this thermodynamic system, has increased in entropy. Energy has spontaneously become more dispersed and spread out in that "universe" than when the glass of ice water was introduced and became a "system" within it.

Entropy and Cosmology

As a finite universe may be considered an isolated system, it may be subject to the Second Law of Thermodynamics, so that its total entropy is constantly increasing. It has been speculated that the universe is fated to a heat death in which all the energy ends up as a homogeneous distribution of thermal energy, so that no more work can be extracted from any source.

If the universe can be considered to have generally increasing entropy, then – as Roger Penrose has pointed out – gravity plays an important role in the increase because gravity causes dispersed matter to accumulate into stars, which collapse eventually into black holes.

Jacob Bekenstein and Stephen Hawking have shown that black holes have the maximum possible entropy of any object of equal size. This makes them likely end points of all entropy-increasing processes, if they are totally effective matter and energy traps. Hawking has, however, recently changed his stance on this aspect.

The role of entropy in cosmology remains a controversial subject. Recent work has cast extensive doubt on the heat death hypothesis and the applicability of any simple thermodynamic model to the universe in general.

Although entropy does increase in the model of an expanding universe, the maximum possible entropy rises much more rapidly – thus entropy density is decreasing with time. This results in an "entropy gap" pushing the system further away from equilibrium. Other complicating factors, such as the energy density of the vacuum and macroscopic quantum effects, are difficult to reconcile with thermo-dynamical models, making any predictions of large-scale thermodynamics extremely difficult.

IRREVERSIBLE PROCESSES

In order that a process may be strictly reversible, it is necessary that the state of the working substance should be one of equilibrium at uniform pressure and temperature throughout.

If heat passes " of itself " from a higher to a lower temperature by conduction, convection or radiation, the transfer cannot be reversed without an expenditure of work. If mechanical work or kinetic energy is directly converted into heat by friction, reversal of the motion does not restore the energy so converted. In all such cases there is necessarily, by Carnot's principle, a loss of efficiency or available energy, accompanied by an increase of entropy, which serves as a convenient measure or criterion of the loss.

A common illustration of an irreversible process is the expansion of a gas into a vacuum or against a pressure less than its own. In this case the work of expansion, pdv, is expended in the first instance in producing kinetic energy of motion of parts of the gas. If this could be co-ordinated and utilized without dissipation, the gas might conceivably be restored to its initial state; but in practice violent local differences of pressure and temperature are produced, the kinetic energy is rapidly converted into heat by viscous eddy friction, and residual differences of temperature are equalized by diffusion throughout the mass.

Even if the expansion is adiabatic, in the sense that it takes place inside a non-conducting enclosure and no heat is supplied from external sources, it will not be isentropic, since the heat supplied by internal friction must be included in reckoning the change of entropy. Assuming that no heat is supplied from external sources and no external work is done, the intrinsic energy remains constant by the first law. The final state of the substance, when equilibrium has been restored, may be deduced from this condition, if the energy can be expressed in terms of the co-ordinates.

But the line of constant energy on – the diagram does not represent the path of the transformation, unless it be supposed to be effected in a series of infinitesimal steps between each of which the substance is restored to an equilibrium state. An irreversible process which permits a more complete experimental investigation is the steady flow of a fluid in a tube.

If the tube is a perfect non-conductor, and if there are no eddies or frictional dissipation, the state of the substance at any point of the tube as to E, p, and v, is represented by the adiabatic or isentropic path, dE= -pdv. The tube varies, the change of kinetic energy of flow, dU, is represented by The flow in this case is reversible, and the state of the fluid is the same at points where the section of the tube is the same.

In practice, however, there is always some frictional dissipation, accompanied by an increase of entropy and by a fall of pressure. In the limiting case of a long fine tube, the bore of which varies in such a manner that U is constant, the state of the substance along a line of flow may be represented by the line of constant total heat, d(E+pv) = o; but in the case of a porous plug or small throttling aperture, the steps of the process cannot be followed, though the final state is the same.

ENTROPY SYSTEM AND QUANTUM MECHANICS

The entropy of a system is defined in terms of the number Ω of accessible microstates consistent with an overall energy in the range E to $E + \delta E$via

$$S = k \ln \Omega.$$

We have already demonstrated that this definition is utterly insensitive to the resolution δE to which the macroscopic energy is measured. In classical mechanics, if a system possesses f degrees of freedom then phase-space is conventionally subdivided into cells of arbitrarily chosen volume h_0^f. The number of accessible microstates is equivalent to the number of these cells in the volume of phase-space consistent with an overall energy of the system lying in the range E to $E + \delta E$. Thus,

$$\Omega = \frac{1}{h_0^f} \int \cdots \int dq_1 \cdots dq_f dp_1 \cdots dp_f,$$

giving

$$S = k \ln\left(\int \cdots \int dq_1 \cdots dq_f dp_1 \cdots dp_f \right) - kf \ln h_0.$$

Thus, in classical mechanics the entropy is undetermined to an arbitrary additive constant which depends on the size of the cells in phase-space. In fact, S increases as the cell size decreases. The second law of thermodynamics is only concerned with changes in entropy, and is, therefore, unaffected by an additive constant. Likewise, macroscopic thermodynamical quantities, such as the temperature and pressure, which can be expressed as partial derivatives of the entropy with respect to various macroscopic parameters are unaffected by such a constant. So, in classical mechanics the entropy is rather like a gravitational potential: it is undetermined to an additive constant, but this does not affect any physical laws. The non-unique value of the entropy comes about because there is no limit to the precision to which the state of a classical system can be specified. In other words, the cell size h_0 can be made arbitrarily small, which corresponds to specifying the particle coordinates and momenta to arbitrary accuracy. However, in quantum mechanics the uncertainty principle sets a definite limit to how accurately the particle coordinates and momenta can be specified. In general,

$$\delta q_i \, \delta p_i \geq h_i$$

where p_i is the momentum conjugate to the generalized coordinate q_i, and δq_i, $\delta p_i$are the uncertainties in these quantities, respectively. In fact, in quantum mechanics the number of accessible quantum states with the overall energy in the range E to $E + \delta E$ is completely determined. This implies that, in reality, the entropy of a system has a unique and unambiguous value. Quantum mechanics can often be "mocked up" in classical mechanics by setting the cell size in phase-space equal to Planck's constant, so that $h_0 = h$. This automatically enforces the most restrictive form of the uncertainty principle, $\delta q_i \delta p_i = h$. In many systems, the substitution $h_0 \rightarrow h$ in Eq. gives the same, unique value for as that obtained from a full quantum mechanical calculation. Consider a simple quantum mechanical system consisting of N non-interacting spinless particles of mass m confined in a cubic box of dimension L. The energy levels of the th particle are given by

$$e_i = \frac{\hbar^2 \pi^2}{2mL^2} \left(n_{il}^2 + n_{i2}^2 + n_{i3}^2 \right),$$

where n_{i1}, n_{i2}, and n_{i3} are three (positive) quantum numbers. The overall energy of the system is the sum of the energies of the individual particles, so that for a general state τ

$$E_r = \sum_{i=1}^{N} e_i .$$

The overall state of the system is completely specified by $3N$ quantum numbers, so the number of degrees of freedom is $f = 3N$. The classical limit corresponds to the situation where all of the quantum numbers are much greater than unity. In this limit, the number of accessible states varies with energy according to our usual estimate $\Omega \propto E^f$.

The lowest possible energy state of the system, the so-called ground-state, corresponds to the situation where all quantum numbers take their lowest possible value, unity. Thus, the ground-state energy E_0 is given by

$$E_0 = \frac{f\hbar^2\pi^2}{2mL^2}.$$

There is only one accessible microstate at the ground-state energy (i.e., that where all quantum numbers are unity), so by our usual definition of entropy

$$S(E_0) = k\ln 1 = 0.$$

In other words, there is no disorder in the system when all the particles are in their ground-states.

Clearly, as the energy approaches the ground-state energy, the number of accessible states becomes far less than the usual classical estimate E^f. This is true for all quantum mechanical systems. In general, the number of microstates varies roughly like

$$\Omega(E) \sim 1 + C(E - E_0)^f ,$$

where C is a positive constant. According to Eq. the temperature varies approximately like

$$T \sim \frac{E - E_0}{kf},$$

provided $\Omega \gg 1$. Thus, as the absolute temperature of a system approaches zero, the internal energy approaches a limiting value E_0(the quantum mechanical ground-state energy), and the entropy approaches the limiting value zero. This proposition is known as the third law of thermodynamics.

At low temperatures, great care must be taken to ensure that equilibrium thermodynamical arguments are applicable, since the rate of attaining equilibrium may be very slow. Another difficulty arises when dealing with a system in which the atoms possess nuclear spins.

Typically, when such a system is brought to a very low temperature the entropy associated with the degrees of freedom not involving nuclear spins becomes negligible. Nevertheless, the number of microstates Ω_s corresponding to the possible nuclear spin orientations may be very large. Indeed, it may be just as large as the number of states at room temperature.

The reason for this is that nuclear magnetic moments are extremely small, and, therefore, have extremely weak mutual interactions. Thus, it only takes a tiny amount of heat energy in the system to completely randomize the spin orientations. Typically, a temperature as small as 10^{-3} degrees kelvin above absolute zero is sufficient to randomize the spins.

Suppose that the system consists of N atoms of spin 1/2. Each spin can have two possible orientations. If there is enough residual heat energy in the system to randomize the spins then each orientation is equally likely. If follows that there are $\Omega_s = 2^N$ accessible spin states. The entropy associated with these states is $S_0 = k \ln \Omega_s = v R \ln 2$. Below some critical temperature, T_0, the interaction between the nuclear spins becomes significant, and the system settles down in some unique quantum mechanical ground-state (e.g., with all spins aligned).

In this situation, $S \rightarrow 0$, in accordance with the third law of thermodynamics. However, for temperatures which are small, but not small enough to "freeze out" the nuclear spin degrees of freedom, the entropy approaches a limiting value S_0 which depends only on the kinds of atomic nuclei in the system. This limiting value is independent of the spatial arrangement of the atoms, or the interactions between them. Thus, for most practical purposes the third law of thermodynamics can be written

as $T \rightarrow 0_+$, $S \rightarrow S_0$,

where 0_+denotes a temperature which is very close to absolute zero, but still much larger than. This modification of the third law is useful because it can be applied at temperatures which are not prohibitively low.

MICROSCOPIC DEFINITION OF ENTROPY (STATISTICAL MECHANICS)

In statistical thermodynamics the entropy is defined as (proportional to) the logarithm of the number of microscopic configurations that result in the observed macroscopic description of the thermodynamic system:

$$S = k_B \ln \Omega$$

where

kB is Boltzmann's constant 1.38066×10–23 J K–1 and

Ωis the number of microstates corresponding to the observed thermodynamic macrostate.

This definition is considered to be the fundamental definition of entropy. In Boltzmann's 1896 Lectures on Gas Theory, he showed that this expression gives a measure of entropy for systems of atoms and molecules in the gas phase, thus providing a measure for the entropy of classical thermodynamics.

In 1877, Boltzmann visualized a probabilistic way to measure the entropy of an ensemble of ideal gas particles, in which he defined entropy to be proportional to the logarithm of the number of microstates such a gas could occupy. Henceforth, the essential problem in statistical thermodynamics, i.e. according to Erwin Schrödinger, has been to determine the distribution of a given amount of energy E over N identical systems.

Statistical mechanics explains entropy as the amount of uncertainty which remains about a system, after its observable macroscopic properties have been taken into account. For a given set of macroscopic variables, like temperature and volume, the entropy measures the degree to which the probability of the system is spread out over different possible quantum states. The more states available to the system with higher probability, the greater the entropy. More specifically, entropy is a logarithmic measure of the density of states. In essence, the most general interpretation of entropy is as a measure of our uncertainty about a system. The equilibrium state of a system maximizes the entropy because we have lost all information about the initial conditions except for the conserved variables; maximizing the entropy maximizes our ignorance about the details of the system. This uncertainty is not of the everyday subjective kind, but rather the uncertainty inherent to the experimental method and interpretative model.

On the molecular scale, the two definitions match up because adding heat to a system, which increases its classical thermodynamic entropy, also increases the system's thermal fluctuations, so giving an increased lack of information about the exact microscopic state of the system, i.e. an increased statistical mechanical entropy.

The interpretative model has a central role in determining entropy. The qualifier "for a given set of macroscopic variables" above has very deep implications: if two observers use different sets of macroscopic variables, then they will observe different entropies. For example, if observer A uses the variables U, V and W, and observer B uses U, V, W, X, then, by changing X, observer B can cause an effect that looks like a violation of the second law of thermodynamics to observer A. In other words: the set of macroscopic variables one chooses must include everything that may change in the experiment, otherwise one might see decreasing entropy.

ENTROPY IN CHEMICAL THERMODYNAMICS

Thermodynamic entropy is central in chemical thermodynamics, enabling changes to be quantified and the outcome of reactions predicted. The second law of thermodynamics states that entropy in the combination of a system and its surroundings increases during all spontaneous chemical and physical processes. Spontaneity in chemistry means "by itself, or without any outside influence", and has nothing to do with speed. The Clausius equation of δqrev/T = ΔS introduces the measurement of entropy change, ΔS. Entropy change describes the direction and quantitates the magnitude of simple changes such as heat transfer between systems – always from hotter to cooler spontaneously. Thus, when a mole of substance at 0 K is warmed by its surroundings to 298 K, the sum of the incremental values of qrev/T constitute each element's or compound's standard molar entropy, a fundamental physical property and an indicator of the amount of energy stored by a substance at 298 K. Entropy change also measures the mixing of substances as a summation of their relative quantities in the final mixture.

Entropy is equally essential in predicting the extent of complex chemical reactions, i.e. whether a process will go as written or proceed in the opposite direction. For such applications, ΔS must be incorporated in an expression that includes both the system and its surroundings, ΔSuniverse = ΔSsurroundings + ΔS system. This expression becomes, via some steps, the Gibbs free energy equation for reactants and products in the system: ΔG [the Gibbs free energy change of the system] = ΔH [the enthalpy change] "T ΔS [the entropy change].

ENTROPY BALANCE EQUATION FOR OPEN SYSTEMS

In chemical engineering, the principles of thermodynamics are commonly applied to "open systems", i.e. those in which heat, work, and mass flow across the system boundary. In a system in which there are flows of both heat ($\dot{Q}$) and work, i.e. $\dot{W}_S$ (shaft work) and P(dV/dt) (pressure-volume work), across the system boundaries, the heat flow, but not the work flow, causes a change in the entropy of the system.

This rate of entropy change is $\dot{Q}/T$ where T is the absolute thermodynamic temperature of the system at the point of the heat flow. If, in addition, there are mass flows across the system boundaries, the total entropy of the system will also change due to this convected flow.

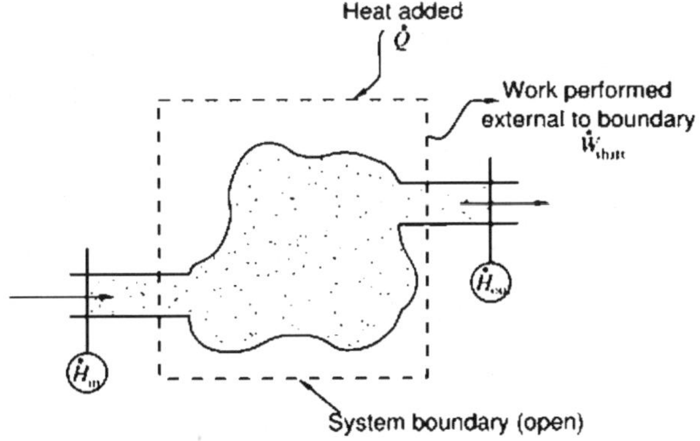

Fig. Entropy Balance Applied to An Open System

To derive a generalized entropy balanced equation, we start with the general balance equation for the change in any extensive quantity Θ in a thermodynamic system, a quantity that may be either conserved, such as energy, or non-conserved, such as entropy. The basic generic balance expression states that dΘ/dt, i.e. the rate of change of Θ in the system, equals the rate at which Θ enters the system at the boundaries, minus the rate at which Θ leaves the system across the system boundaries, plus the rate at which Θ is generated within the system.

Using this generic balance equation, with respect to the rate of change with time of the extensive quantity entropy S, the entropy balance equation for an open thermodynamic system is:

$$\frac{dS}{dt} = \sum_{k=1}^{K} \dot{M}_k \hat{S}_k + \frac{\dot{Q}}{T} + \dot{S}_{gen}$$

where

$\sum_{k=1}^{K} \dot{M}_k \hat{S}_k$ = the net rate of entropy flow due to the flows of mass into and out of the system (where $\hat{S}$ = entropy per unit mass).

$\frac{\dot{Q}}{T}$ = the rate of entropy flow due to the flow of heat across the system boundary.

$\dot{S}_{gen}$ = the rate of internal generation of entropy within the system.

Note, also, that if there are multiple heat flows, the term $\dot{Q}/T$ is to be replaced by $\sum \dot{Q}_j / T_j$, where $\dot{Q}_j$ is the heat flow and Tj is the temperature at the jth heat flow port into the system.

6

Laws of Thermodynamics

INTRODUCTION

Thermal Equilibrium: A system with many microscopic components (for example, a gas, a liquid, a solid with many molecules) that is isolated from all forms of energy exchange and left alone for a "long time" moves toward a state of thermal equilibrium. A system in thermal equilibrium is characterized by a set of macroscopic quantities that depend on the system in question and characterize its "state" (such as pressure, volume, density) that do not change in time. Two systems are said to be in (mutual) thermal equilibrium if, when they are placed in "thermal contact" (basically, contact that permits the exchange of energy between them), their state variables do not change. A system is said to be in thermal equilibrium when its temperature is stable (i.e does not change over time). Mathematically, this could be expressed as:

$$\frac{dT}{dt} = 0.$$

ZEROTH LAW OF THERMODYNAMICS

If system A is in thermal equilibrium with system C, and system B is in thermal equilibrium with system C, then system A is in thermal equilibrium with system B.

Temperature and Thermometers: The point of the Zeroth Law is that it is the basis of the thermometer. A thermometer is a portable device whose thermal state is related linearly to some simple property, for example its density or pressure. Once a suitable temperature scale is defined for the device, one can use it to measure the temperature of a variety of disparate systems in thermal equilibrium. Temperature thus characterizes thermal equilibrium.

Temperature Scales:

- *Fahrenheit:* This is one of the oldest scales, and is based on the coldest temperature that could be achieved with a mix of ice and alcohol. In it the freezing point of water is at 32° F, the boiling point of water is at 212° F.

- *Celsius or Centigrade:* This is a very sane system, where the freezing point of water is at 0° C and the boiling point is at 100° C. The degree size is thus as big 9/5 as the Fahrenheit degree.
- *Kelvin or Absolute:* 0°K is the lowest possible temperature, where the internal energy of a system is at its absolute minimum. The degree size is the same as that of the Centigrade or Celsius scale. This makes the freezing point of water at atmospheric pressure 273.16°K, the boiling point at 373.16°K.
- Thermal Expansion

$$\Delta L = \alpha L \Delta T$$

- where α is the coefficient of linear expansion. If one applies this in three dimensions:

$$\Delta V = \beta V \Delta T$$

- where $\beta = 3\alpha$.
- Ideal Gas Law

$$PV = nRT = NkT$$

- where J/mol-K, and $k = R/N_A = 1.38 \times 10^{-23}$ J/K.

THERMAL EQUILIBRIUM BETWEEN MANY SYSTEMS

Many systems are said to be in equilibrium if the small exchanges (due to Brownian motion, for example) between them do not lead to a net change in the total energy summed over all systems.

Consider N systems in adiabatic isolation from the rest of the universe (i.e no heat exchange is possible outside of these N systems), all of which have a constant volume and composition, who can only exchange heat with one another. The combined first and second laws relate the fluctuations in total energy äU to the temperature of the ith system Ti and the entropy fluctuation in the ith system äSi by,

$$\delta U = \sum_{i}^{N} T_i \delta S_i .$$

The adiabatic isolation of the system from the remaining universe requires that the total sum of the entropy fluctuations vanishes,

$$\sum_{i}^{N} \delta S_i = 0 ,$$

that is, entropy can only be exchanged between the N systems. This constraint can be used to re-arrange the expression for the total energy fluctuation to give,

$$\delta U = \sum_{i}^{N} (T_i - T_j)\, \delta S_i ,$$

where Tj is the temperature of any system j we may choose to single out among the N systems. Finally, equilibrium requires the total fluctuation in energy to vanish, so we arrive at,

$$\sum_{i}^{N}(T_i - T_j)\,\delta S_i = 0$$

which can be thought of as the vanishing of the product of an anti-symmetric matrix Ti " Tj and a vector of entropy fluctuations äSi. In order for a non-trivial solution to exist,

$$\delta S_i \neq 0,$$

the determinant of the matrix formed by Ti " Tj must vanish for all choices of N. However, according to Jacobi's theorem, the determinant of a NxN anti-symmetric matrix is always zero if N is odd, although for N even we find that all of the entries must vanish, Ti " Tj = 0, in order to obtain a vanishing determinant, and hence Ti = Tj at equilibrium. This non-intuitive result means that an odd number of systems are always in equilibrium regardless of their temperatures and entropy fluctuations, while equality of temperatures is only required between an even number of systems to achieve equilibrium in the presence of entropy fluctuations.

The zeroth law solves this odd vs. even paradox, because it can readily be used to reduce an odd-numbered system to an even number by considering any three of the N systems and eliminating one by application of its principle, and hence reduce the problem to even N which subsequently leads to the same equilibrium condition that we expect in every case, i.e., Ti = Tj. The same result applies to fluctuations in any extensive quantity, such as volume (yielding the equal pressure condition), or fluctuations in mass (leading to equality of chemical potentials), and therefore the zeroth law carries implications for a great deal more than temperature alone. In general, the zeroth law breaks a certain kind of anti-symmetry that exists in the first and second laws.

Temperature and the Zeroth Law

In the space of thermodynamic parameters, zones of constant temperature will form a surface, which provides a natural order of nearby surfaces. It is then simple to construct a global temperature function that provides a continuous ordering of states. Note that the dimensionality of a surface of constant temperature is one less than the number of thermodynamic parameters (thus, for an ideal gas described with 3 thermodynamic parameter P, V and n, they are 2D surfaces). The temperature so defined may indeed not look like the Celsius temperature scale, but it is a temperature function.

For example, if two systems of ideal gas are in equilibrium, then P1V1/N1 = P2V2/N2 where Pi is the pressure in the ith system, Vi is the volume, and Ni is the 'amount' (in moles, or simply number of atoms) of gas.

The surface PV / N = const defines surfaces of equal temperature, and the obvious (but not only) way to label them is to define T so that PV / N = RT where R is some constant. These systems can now be used as a thermometer to calibrate other systems.

LAWS OF CONSERVATION OF ENERGY

The principle of conservation of energy states that energy cannot be created or destroyed, but can be changed from one form to another.

There are many ways in which the principle of conservation of energy may be stated, depending on the application. In special form the principle of conservation of mechanical energy states that the mechanical energy of any system of bodies connected together in any way is conserved.

The subject of thermodynamics is founded on four experience, which are called as the laws of thermodynamics. Each law explains a particular constraint on the properties of thc world.

The law of conservation of energy states that energy is neither created nor destroyed. Therefore the sum of all the energies in the system is a constant.

Consider a pendulum as shown in figure:

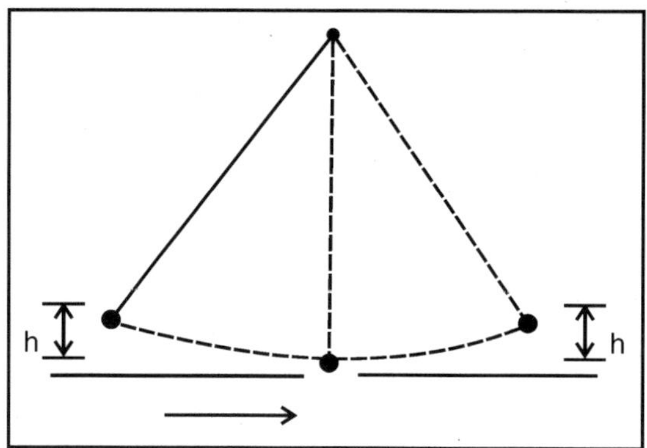

The formula to calculate the potential energy is:

PE = mgh

The mass of the ball = 10kg

The height, h = 0.2m

The acceleration due to gravity, g = 9.8 m/s^2

Substitute the values into the formula and you get:

PE = 19.6J (J = Joules, unit of energy)

The position of the blue ball is PE = 19.6J while the Kinetic Energy (KE) = 0.

As the blue ball approaches the purple ball position, the PE decreases and KE is increases. At the middle position the blue and purple ball is given as PE = KE.. The position of the purple ball where the Kinetic Energy is at maximum and the Potential Energy (PE) = 0. At this point, all the PE has transformed into KE. Therefore now the KE = 19.6J while the PE = 0. The position of the pink ball is where the Potential Energy (PE) is once again at its maximum and the Kinetic Energy (KE) = 0.

We can say that:

PE + KE = 0

PE = -KE

The sum of PE and KE is the total mechanical energy:

Total Mechanical Energy = PE + KE

FIRST LAW OF THERMODYNAMICS

The first law of thermodynamics is the application of the conservation of energy principle to heat and thermodynamic processes:

The change in internal energy of a system is equal to the heat added to the system minus the work done by the system.

$$\Delta U = Q - W$$

Change in internal energy | Heat added to the system | Work done by the system

The first law makes use of the key concepts of internal energy, heat, and system work. It is used extensively in the discussion of heat engines.

It is typical for chemistry texts to write the first law as $\Delta U = Q + W$. It is the same law, of course - the thermodynamic expression of the conservation of energy principle. It is just that W is defined as the work done on the system instead of work done by the system. In the context of physics, the common scenario is one of adding heat to a volume of gas and using the expansion of that gas to do work, as in the pushing down of a piston in an internal combustion engine. In the context of chemical reactions and process, it may be more common to deal with situations where work is done on the system rather than by it.

ENTHALPY

Four quantities called "thermodynamic potentials" are useful in the chemical thermodynamics of reactions and non-cyclic processes. They are internal energy, the enthalpy, the Helmholtz free energy and the Gibbs free energy. Enthalpy is defined by

$$H = U + PV$$

where P and V are the pressure and volume, and U is internal energy. Enthalpy is then a precisely measurable state variable, since it is defined in terms of three other precisely definable state variables. It is somewhat parallel to the first law of thermodynamics for a constant pressure system

$$Q = \Delta U + P\Delta V \text{ since in this case } Q = \Delta H$$

It is a useful quantity for tracking chemical reactions. If as a result of an exothermic reaction some energy is released to a system, it has to show up in some measurable form in terms of the state variables. An increase in the enthalpy $H = U + PV$ might be associated with an increase in internal energy which could be measured by calorimetry, or with work done by the system, or a combination of the two.

The internal energy U might be thought of as the energy required to create a system in the absence of changes in temperature or volume.

But if the process changes the volume, as in a chemical reaction which produces a gaseous product, then work must be done to produce the change in volume. For a constant pressure process the work you must do to produce a volume change ÄV is PÄV. Then the term PV can be interpreted as the work you must do to "create room" for the system if you presume it started at zero volume.

SYSTEM WORK

When work is done by a thermodynamic system, it is ususlly a gas that is doing the work. The work done by a gas at constant pressure is:

$$W = P\Delta V$$

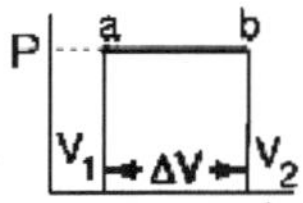

The line from a to b represents an expansion of a gas at constant pressure. The work done is the area under the curve.

For non-constant pressure, the work can be visualized as the area under the pressure-volume curve which represents the process taking place. The more general expression for work done is:

The integral expression gives the exact area under the curve which is equal to the work.

$$W = \int_{V_1}^{V_2} P\,dV$$

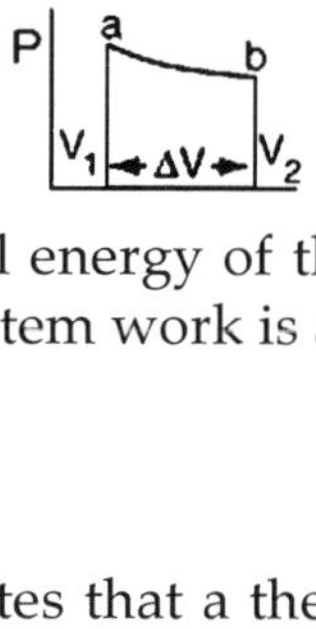

Work done by a system decreases the internal energy of the system, as indicated in the First Law of Thermodynamics. System work is a major focus in the discussion of heat engines.

DESCRIPTION

The first law of thermodynamics basically states that a thermodynamic system can store or hold energy and that this internal energy is conserved. Heat is a process by which energy is added to a system from a high-temperature source, or lost to a low-temperature sink. In addition, energy may be lost by the system when it does mechanical work on its surroundings, or conversely, it may gain energy as a result of work done on it by its surroundings. The first law states that this energy is conserved: The change in the internal energy is equal to the amount added by heating minus the amount lost by doing work on the environment. The first law can be stated mathematically as:

$$dU = dQ - \delta W$$

where dU is a small increase in the internal energy of the system, δQ is a small amount of heat added to the system, and äW is a small amount of work done by the system.

The δ's before the heat and work terms are used to indicate that they describe an increment of energy which is to be interpreted somewhat differently than the dU increment of internal energy. Work and heat are processes which add or subtract energy, while the internal energy U is a particular form of energy associated with the system. Thus the term "heat energy" for δQ means "that amount of energy added as the result of heating" rather than referring to a particular form of energy.

Likewise, the term "work energy" for δw means "that amount of energy lost as the result of work". The most significant result of this distinction is the fact that one can clearly state the amount of internal energy possessed by a thermodynamic system, but one cannot tell how much energy has flowed into or out of the system as a result of its being heated or cooled, nor as the result of work being performed on or by the system.

The first explicit statement of the first law of thermodynamics was given by Rudolf Clausius in 1850: "There is a state function E, called 'energy', whose differential equals the work exchanged with the surroundings during an adiabatic process." Note that the above formulation is favored by engineers and physicists. Chemists prefer a second form, in which the work term δw is defined as the work done on the system, and therefore insert a plus sign in the above equation before the work term. This article will use the first definition exclusively.

APPLICATION OF FIRST LAW OF THERMODYNAMICS

Constant volume

Consider a system as shown below

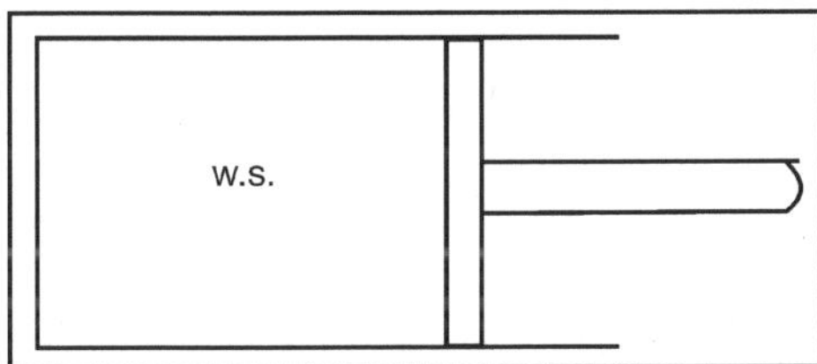

In the above figure, the Work output = 0 and Constant Volume Processes are invariably. All terms in the above equation are eliminated with the exception of:-

$$E_1 + Q_S = E_2$$

i.e. all the heat supplied goes to increasing the Internal Energy of the WS

CONSTANT PRESSURE

Consider a system:
Non-Flow system
Force on the Piston = P A
Work Done = P A ι = $P(V_2 - V_1)$

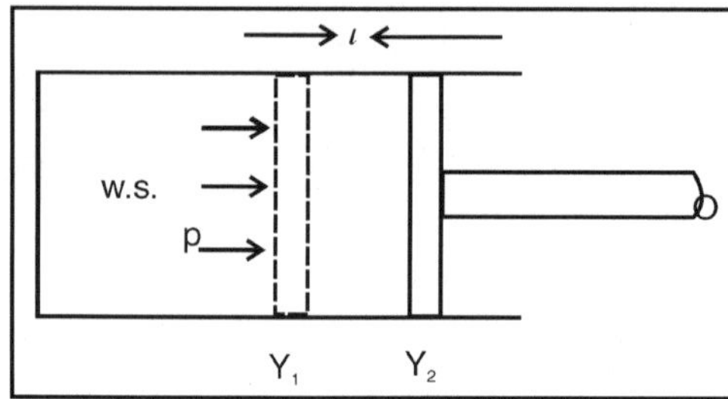

In the First Law equation, most terms are eliminated and left with:-

$$E_1 + Q_S = E_2 + W$$

$$\therefore\ QS = (E_2 - E_1) + P(V_2 - V_1) = (E_2 + P_2V_2) - (E_1 + P_1V_1)$$

$$= H_2 - H_1$$

i.e. The Heat Supplied = The Gain in ENTHALPY.

ADIABATIC PROCESS

An adiabatic process is one that occurs without the exchange of heat with the surroundings.

If the gas-piston system were insulated so that heat could not get in or out, any expansion or compression would occur adiabatically.

This is a Process during which there is NO heat transfer of heat between the WS and the surroundings.

For Non Flow system:

$$E_1 = E_2 + W$$

Work Done = $(E_1 - E_2)$ i.e. Loss of Internal Energy

Therefore Work Done = Loss in Enthalpy.

Figure below shows P-V diagrams for these two processes.

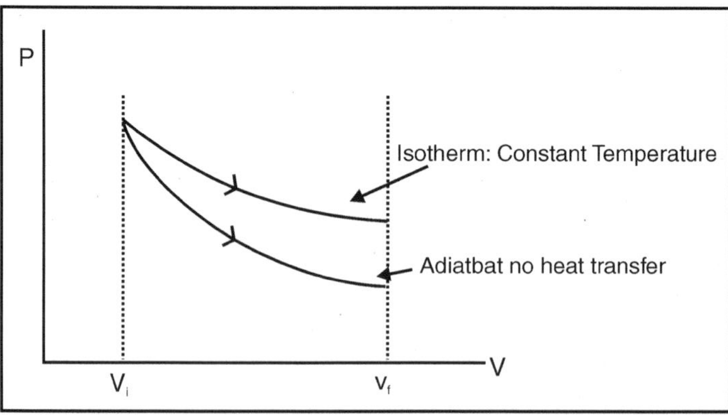

This figure compares two processes that begin with the same state and involve expansion to the same volume. For the isothermal process, the product of $P \cdot V$ remains constant since T remains constant. Since the temperature decrease in an adiabatic process, therefore the pressure is also less.

STEADY FLOW ENERGY EQUATION

The steady flow energy equation relates to open systems working under steady conditions i.e in which conditions do not change with time.

The boundary encloses a system through which fluid flows at a constant rate, and heat transfer occurs and external work is done all under steady conditions.

The equation for steady flow is written per unit mass as

$$q - w = \Delta\left(h + \frac{1}{2}v^2 + gz\right)$$

q = heat transfer across boundary per unit mass
w = external work done by system per unit mass
z = fluid height
v = fluid velocity
h = fluid enthalpy (u (internal energy + pv (pressure. specific volume)
Consider a system as shown in figure below:

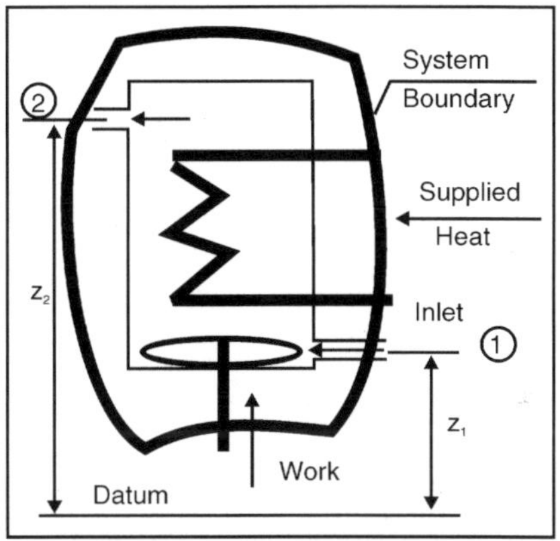

By applying first law of thermodynamics to above system, we can write equation as:

$$dQ/dt + dW/dt = m[\Delta h + \Delta C^2/2 + \Delta Zg]$$

Where,
dQ/dt = Supplied heat to the system per unit time,
dW/dt = Input work to the system per unit time,
m = Mass flow rate,
Δh = h2-h1
h = Specific enthalpy,
$\Delta C^2/2$ = Difference in kinetic energy between outlet and inlet,
Z = Height measured from some reference datum,
1, 2 = refer to inlet and outlet, respectively.

STEADY FLOW ENERGY EQUATION OF GAS TURBINE

Consider an example of a turbine as shown in figure 4.3 below. The engine is designed to produce about 84,000 lbs of thrust at takeoff.

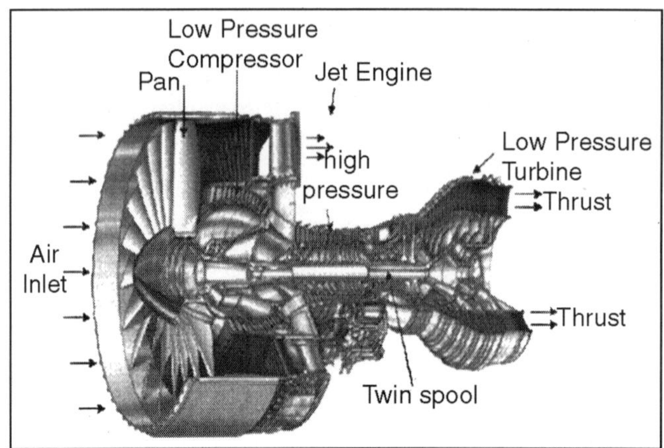

The fan and low pressure compressor are driven by the low pressure turbine. The high pressure compressor is driven by the high pressure turbine.

f = total pressure ratio across the fan Å 1.4

c = total pressure ratio across the fan + compressor Å 45

Tin let = 300K

Heat transfer from the gas streams is negligible so we write the First Law equation as:

$$Q - W_S = m(h_{T2} - h_{T1})$$

Now we see that:

$$-W_S = m_f \Delta h_{Tf} + m_c \Delta h_{Tc}$$
$$= m_f c_p \Delta T_{Tf} + m_c c_p \Delta T_{Tc}$$

In above equations we obtain the temperature change by assuming that the compression process is quasi-static and adiabatic.

So

$$\frac{T_2}{T_1} = \left(\frac{P_2}{P_1}\right)^{\frac{\gamma-1}{\gamma}}$$

then

$$\left(\frac{T_{T_2}}{T_{T_1}}\right)_{fan} = (\pi_f)^{\frac{\gamma-1}{\gamma}} = 1.1 \Rightarrow \Delta T_{T_{fan}} = 30K$$

$$\left(\frac{T_{T_2}}{T_{T_1}}\right)_{core} = (\pi_{core})^{\frac{\gamma-1}{\gamma}} = 3.0 \Rightarrow \Delta T_{T_{core}} = 600K$$

$$-W_S = 610\frac{Kg}{s} \cdot 30K \cdot 1008\frac{J}{KgK} + 120\frac{Kg}{s} \cdot 600K \cdot 1008\frac{J}{KgK}$$

= 91 x 106 Joules/sec

W_s = -91 Megawatts (Negative sign implies work doneon the fluid)

Note that 1 Hp = 745 watts

STEADY FLOW ENERGY EQUATION OF NOZZLE

Consider a rocket with a chamber and nozzle. In this the liquid propellants moves inside the chamber, which converts the chemical energy into thermal energy.

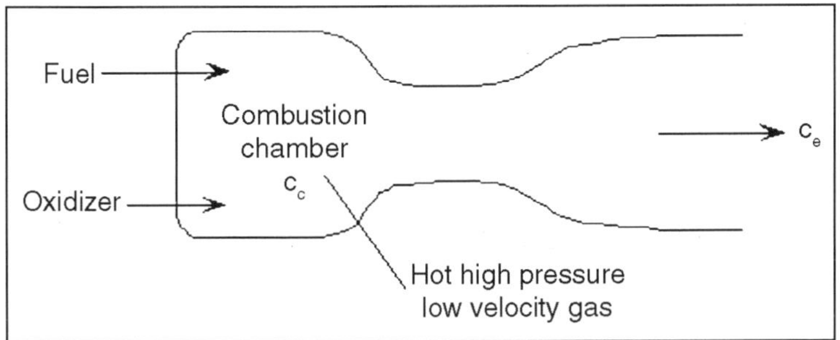

Once the rocket is operating we see that all the flow processes are steady, hence we can use the steady flow energy equation.

We assume that the gas behaves as an ideal gas, and there will be no external work.

Then we can write the First Law as

$$q_{1-2} - W_{S1-2} = h_{T2} - h_{T1}$$

which becomes $h_{T2} = h_{T1}$
or

$$C_p T_c + \frac{C_c^2}{2} = C_p T_e + \frac{C_e^2}{2}$$

therefore

$$C_e = \sqrt{2C_p(T_c - T_e)}$$

If we assume quasi-static, adiabatic expansion then

$$\frac{T_e}{T_c} = \left(\frac{P_e}{P_c}\right)^{\frac{\gamma-1}{\gamma}}$$

so

$$C_e = \sqrt{2C_p T_c \left[1 - \left(\frac{P_e}{P_c}\right)^{\frac{\gamma-1}{\gamma}}\right]}$$

Where Tc and pc are conditions in the combustion chamber and pe is the external static pressure.

STEADY FLOW ENERGY EQUATION OF COMPRESSORS

An axial compressor is typically made up of many alternating rows of rotating and blades called as rotors and stators respectively as shown in Figures. The first row is typically called as the inlet guide vanes or IGV. Each successive rotor-stator pair is called a compressor stage.

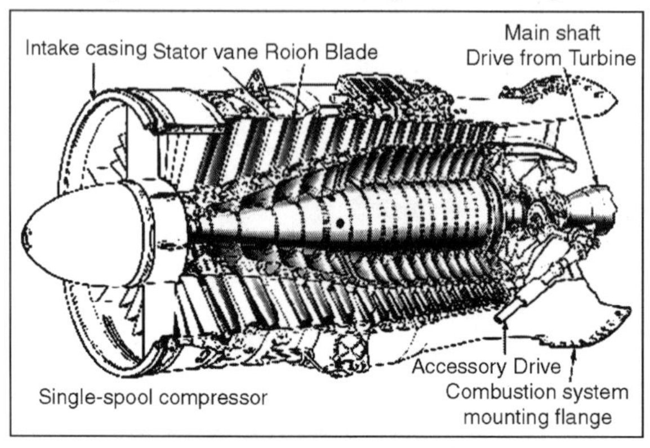

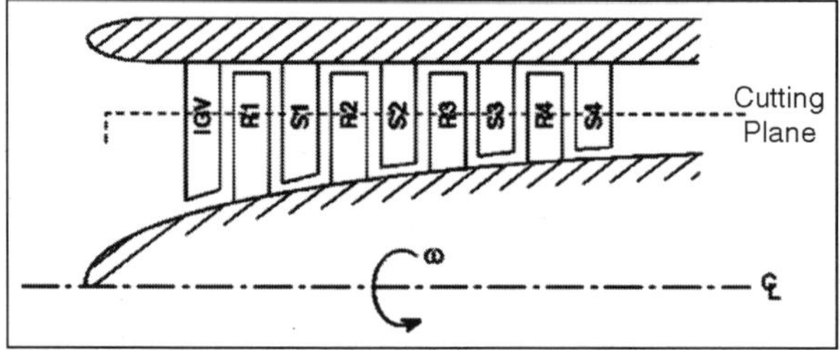

Considering Bernoulli Equation, where P_T is the stagnation pressure, p is the static pressure and velocity (u is radial, v is tangential, w is axial).

$$P_T = p + \frac{1}{2}p(u^2 + v^2 + w^2)$$

In this the rotor increases the energy and angular momentum by adding to the kinetic energy as $1/2rv^2$. We see that a compressor look like as shown in Figure

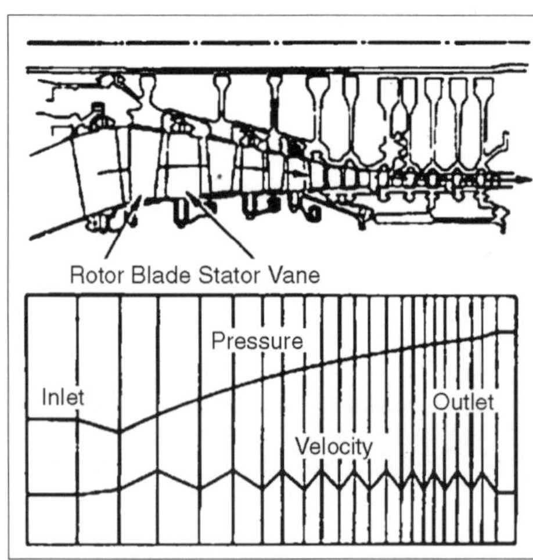

Note that the IGV adds no energy to the flow. It is designed to add swirl in the direction of motion to lower the Mach number of the flow.

HEAT SOURCE

A heat source is anything that can heat up a spacecraft.

Heat sources can be external or internal. External heat sources include:

- The Sun
- Reflected sunlight from planets and moons
- Heating by friction when traveling through an atmosphere or gas clouds, and
- Released heat from planets.

Heat source such as engine, stove or melting pot, will generate a vertical air flow. See figure.

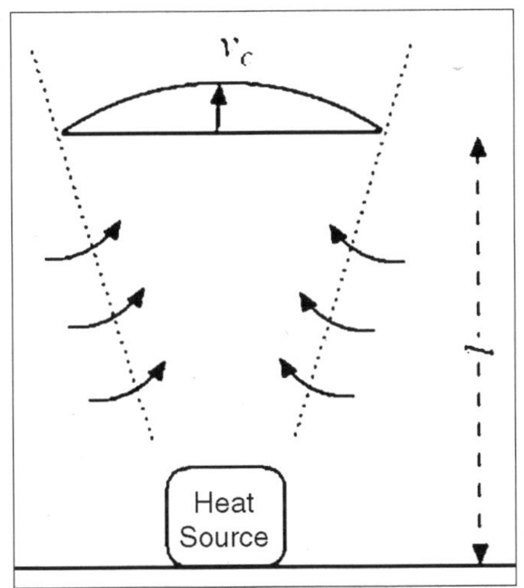

Air Velocity

The air velocity is the center of the air flow which is at a distance above the floor and can be written as

$$v_c = c_1 \, (1000 \, P / l)^{1/3} \quad (1)$$

where

v_c = air velocity in the center of the air flow (m/s)

c_1 = constant characterizing the actual application, typical values between 1 to 2.

P = heating power from the source (W)

l = distance above the floor and the heat source (m)

Air Flow Volume

The air flow in a distance above the floor can be calculated as

$$Q = c_2 \, P^{1/3} \, l^{5/3} \quad (2)$$

where

Q = air flow volume (m^3/s)

c_2 = *constant characterizing the actual application, typical values between 0.05 to 0.15*

HEAT SINK

Heat sinks are cooling mechanisms used to draw out thermal energy from a variety of electronic components. Heat Sink prevent components from overheating. The most common heat sink applications are for computer CPU's, microprocessor chips and circuit boards.

The materials used to construct a heat sink are aluminum and copper, because of its high conductivity. Gold plating is also heat sinks, as it is used to increase the transfer of thermal energy. Depending on design, a fan can be used to make heat sink airflow that benefits the cooling process.

SECOND LAW OF THERMODYNAMICS

The second law of thermodynamics is an expression of the universal law of increasing entropy, stating that the entropy of an isolated system which is not in equilibrium will tend to increase over time, approaching a maximum value at equilibrium. The second law traces its origin to French physicist Sadi Carnot's 1824 paper Reflections on the Motive Power of Fire, which presented the view that motive power (work) is due to the fall of caloric (heat) from a hot to cold body (working substance). In simple terms, the second law is an expression of the fact that over time, ignoring the effects of self-gravity, differences in temperature, pressure, and density tend to even out in a physical system that is isolated from the outside world. Entropy is a measure of how far along this evening-out process has progressed.

There are many versions of the second law, but they all have the same effect, which is to explain the phenomenon of irreversibility in nature.

VERSIONS OF THE LAW

There are many ways of stating the second law of thermodynamics, but all are equivalent in the sense that each form of the second law logically implies every other form. Thus, the theorems of thermodynamics can be proved using any form of the second law and third law The formulation of the second law that refers to entropy directly is as follows:

In a system, a process can occur only if it increases the total entropy of the universe. Thus, while a system can undergo some physical process that decreases its own entropy, the entropy of the universe (which includes the system and its surroundings) must increase overall. (An exception to this rule is a reversible or "isentropic" process, such as frictionless adiabatic compression.) Processes that decrease total entropy of the universe are impossible. If a system is at equilibrium, by definition no spontaneous processes occur, and therefore the system is at maximum entropy. Also, due to Rudolf Clausius, is the simplest formulation of the second law, the heat formulation or Clausius statement: Heat cannot spontaneously flow from a material at lower temperature to a material at higher temperature.

Informally, "Heat doesn't flow from cold to hot (without work input)", which is obviously true from everyday experience. For example in a refrigerator, heat flows from cold to hot, but only when aided by an external agent, i.e. the compressor. Note that from the mathematical definition of entropy, a process in which heat flows from cold to hot has decreasing entropy. This is allowable in a non-isolated system, however only if entropy is created elsewhere, such that the total entropy is constant or increasing, as required by the second law. For example, the electrical energy going into a refrigerator is converted to heat and goes out the back, representing a net increase in entropy.

A third formulation of the second law, by Lord Kelvin, is the heat engine formulation, or Kelvin statement: It is impossible to convert heat completely into work. That is, it is impossible to extract energy by heat from a high-temperature energy source and then convert all of the energy into work. At least some of the energy must be passed on to heat a low-temperature energy sink. Thus, a heat engine with 100% efficiency is thermodynamically impossible.

Microscopic Systems

Thermodynamics is a theory of macroscopic systems at equilibrium and therefore the second law applies only to macroscopic systems with well-defined temperatures. On scales of a few atoms, the second law does not apply; for example, in a system of two molecules, it is possible for the slower-moving ("cold") molecule to transfer energy to the faster-moving ("hot") molecule. Such tiny systems are outside the domain of classical thermodynamics, but they can be investigated in quantum thermodynamics by using statistical mechanics. For any isolated system with a mass of more than a few picograms, the second law is true to within a few parts in a million.

Energy Dispersal

The second law of thermodynamics is an axiom of thermodynamics concerning heat, entropy, and the direction in which thermodynamic processes can occur. For example, the second law implies that heat does not spontaneously flow from a cold material to a hot material, but it allows heat to flow from a hot material to a cold material. Roughly speaking, the second law says that in an isolated system, concentrated energy disperses over time, and consequently less concentrated energy is available to do useful work. Energy dispersal also means that differences in temperature, pressure, and density even out.

Again roughly speaking, thermodynamic entropy is a measure of energy dispersal, and so the second law is closely connected with the concept of entropy. The second law says that temperature differences between systems in contact with each other tend to even out and that work can be obtained from these non-equilibrium differences, but that loss of heat occurs, in the form of entropy, when work is done. Pressure differences, density differences, and particularly temperature differences, all tend to equalize if given the opportunity. This means that an isolated system will eventually come to have a uniform temperature. A heat engine is a mechanical device that provides useful work from the difference in temperature of two bodies:

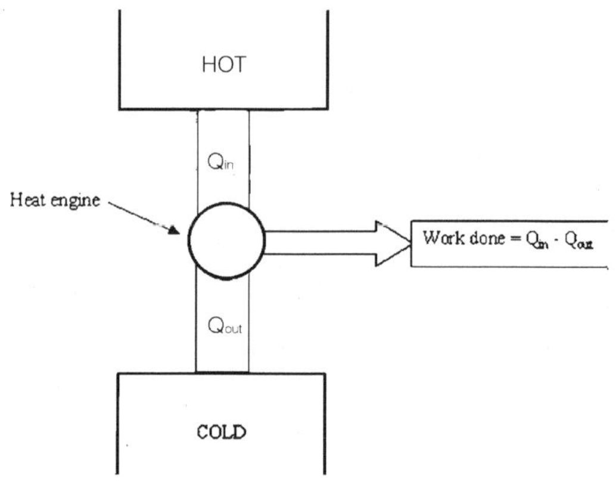

Fig. Heat Engine Diagram

During the 19th century, the second law was synthesized, essentially, by studying the dynamics of the Carnot heat engine in coordination with James Joule's Mechanical equivalent of heat experiments. Since any thermodynamic engine requires such a temperature difference, it follows that no useful work can be derived from an isolated system in equilibrium; there must always be an external energy source and a cold sink. By definition, perpetual motion machines of the second kind would have to violate the second law to function. The first theory on the conversion of heat into mechanical work is due to Nicolas Léonard Sadi Carnot in 1824. He was the first to realize correctly that the efficiency of this conversion depends on the difference of temperature between an engine and its environment.

Recognizing the significance of James Prescott Joule's work on the conservation of energy, Rudolf Clausius was the first to formulate the second law in 1850, in this form: heat does not spontaneously flow from cold to hot bodies. While common knowledge now, this was contrary to the caloric theory of heat popular at the time, which considered heat as a liquid. From there he was able to infer the law of Sadi Carnot and the definition of entropy.

Established in the 19th century, the Kelvin-Planck statement of the Second Law says, "It is impossible for any device that operates on a cycle to receive heat from a single reservoir and produce a net amount of work." This was shown to be equivalent to the statement of Clausius.

The Ergodic hypothesis is also important for the Boltzmann approach. It says that, over long periods of time, the time spent in some region of the phase space of microstates with the same energy is proportional to the volume of this region, i.e. that all accessible microstates are equally probable over long period of time. Equivalently, it says that time average and average over the statistical ensemble are the same.

Using quantum mechanics it has been shown that the local von Neumann entropy is at its maximum value with an extremely high probability, thus proving the second law. The result is valid for a large class of isolated quantum systems (e.g. a gas in a container). While the full system is pure and has therefore no entropy, the entanglement between gas and container gives rise to an increase of the local entropy of the gas. This result is one of the most important achievements of quantum thermodynamics.

Informal Descriptions

The second law can be stated in various succinct ways, including:

- It is impossible to produce work in the surroundings using a cyclic process connected to a single heat reservoir (Kelvin, 1851).
- It is impossible to carry out a cyclic process using an engine connected to two heat reservoirs that will have as its only effect the transfer of a quantity of heat from the low-temperature reservoir to the high-temperature reservoir (Clausius, 1854).
- If thermodynamic work is to be done at a finite rate, free energy must be expended.

KELVIN PLANCK'S STATEMENT

Kelvin-Planck's statement is based on the fact that the efficiency of the heat engine cycle is never 100%. This means that in the heat engine some heat is always rejected to lower the temperature. The heat engine cycle always operates between two heat reservoirs and produces work.

The statement made by Kelvin-Planck for third law of thermodynamics says, "It is impossible for a heat engine to produce net work in a complete cycle if it exchanges heat only with bodies at a single fixed temperature."

The Kelvin-Planck statement tells the condition for producing work within the cycle, the Carnot's statement tells maximum work or efficiency that can be obtained within the cycle.

CLASSIUS STATEMENT

It is impossible to construct a refrigerator operating in a cycle whose sole effect is, to transfer heat from a cooler object to a hotter one.

Heat always flows from high temperature to low temperature as shown in figure.

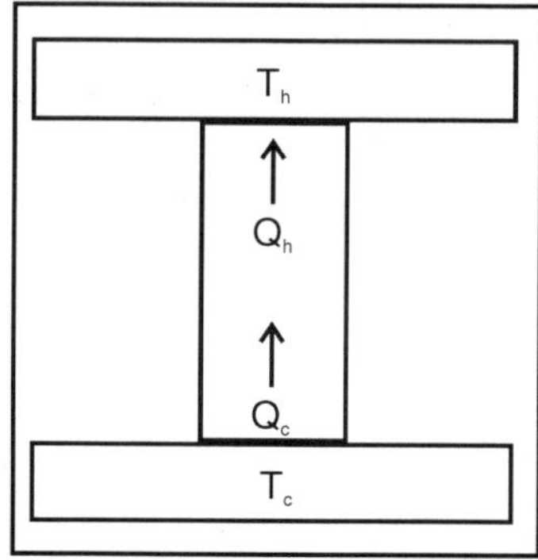

As we know from earlier statement that the natural tendency of the heat is to flow from high temperature to low temperature. The statement of second law of thermodynamics explained by Clausius is that "It is impossible to construct a device which is operating in a cycle, will produce no effect other than the transfer of heat from a colder to a hotter body."

EQUIVALENCE OF STATEMENTS

The equivalence of Clausius and Kelvin plank statements is demonstrated as shown in fig. In this the violation of each statement implies the violation of the other. Here we show the violation of Clausius statement which implies a violation of Kelvin-Plank statement as shown in figure:

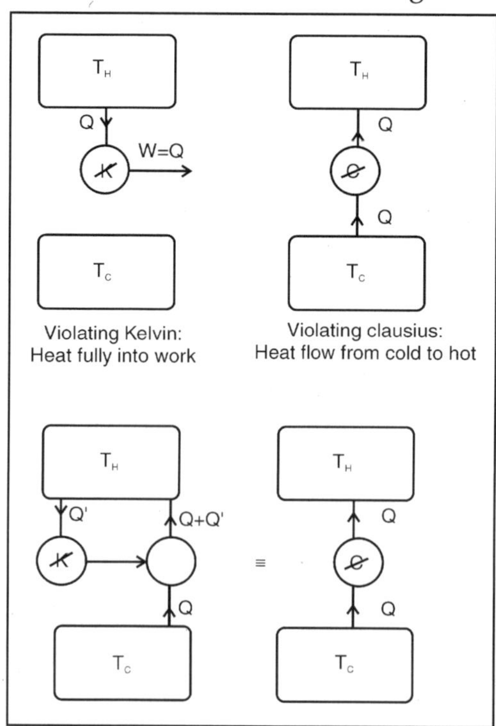

In figure, the top line shows two yellow engines representing engines violations. The second line shows that an engine which violates Kelvin's statement with a heat pump violates Clausius's statement also. A similar drawing can be used to show that a pump which violates Clausius's statement, together with a normal heat engine.

Consider a System shown in figure shown below:

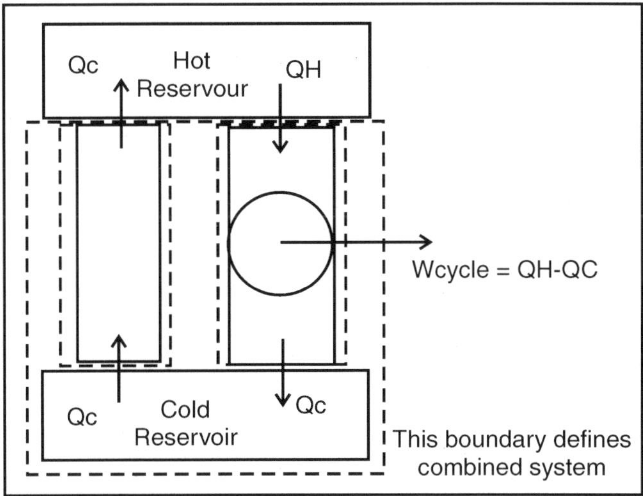

Fig. Shows two systems working between cold and hot reservoir.

The left side of the system transfers energy Qc from the cold reservoir to the hot reservoir by heat transfer without effecting the occurring and Clausius statement. The system on the right operates in a cycle while receiving QH (>Qc) from the hot reservoir and rejecting Qc to the cold reservoir. This delivers work Wcycle to the surroundings. The energy flows are labeled in the figure.

The above figure consists of cold reservoir and two systems. This system can be regarded as executing cycle as one part undergoes a cycle and the other two parts experience no net change. The system receives energy (QH-Qc) by heat transfer from a single reservoir, the hot reservoir, and produces an equivalent amount of work. Thus the violation of the clausius statement implies the violation of Kelvin-Planck statement. In the same way it can be shown that the violation Kelvin-Planck statement implies the violation of Clausius statement.

PERPETUAL-MOTION MACHINE

Perpetual motion machines are devices which operate under the principle of perpetual motion. It means that the machine will continue to perform a function repeatedly without stopping or requiring any type of human interaction. The perpetual motion machine creates energy from nothing and converts energy into work in some manner. Often, the idea of the water screw or siphon is held up as an example of a perpetual motion machine. Perpetual-motion machine is a type of device that would be able to operate continuously and supply useful work, in violation of the laws of thermodynamics.

It produce more energy in the form of work than is supplied to it in the form of heat violating the first law of thermodynamics, which is a special case of the law of conservation of energy and is known as a perpetual-motion machine of the first kind.

A machine that would completely convert heat from a warm body into work, without letting any heat flow into a cooler body, would violate the second law of thermodynamics, which is concerned with entropy changes, and is known as a perpetual-motion machine of the second kind.

PERPETUAL-MOTION MACHINE OF FIRST KIND

A Perpetual Motion Machine of the First Kind is a mechanism which, once in motion, continues to do useful work without an input of energy and produces more energy than is absorbed in its operation.

A machine has a supply of energy. As the machine operates it does only one of two things. Either it does the work or it produces waste heat. The sum total of the original energy, the work done, and the waste heat have to equal a constant.

Perpetual motion machines of the first kind violate the idea of energy conservation. Eg: Perpetual Lamp. This lamp produces heat or light, but never drains an energy to do so.

PERPETUAL-MOTION MACHINE OF SECOND KIND

A Perpetual Motion Machine of the Second Kind is a device that extracts heat from a source and then converts this heat into other forms of energy. The first law demands that all machines have a source of energy, but, it doesn't limit how much of this energy a machine can use for work.

Perpetual motion machines of the second kind operate by extracting energy at some point in their cycle and use this energy for work. It is possible to decrease entropy over a portion of a machine cycle or in a part of a machine, but the remaining cycle produces more than enough entropy to make up for it. For example, it is possible to decrease entropy by cooling the innards of a refrigerator, but only by producing much more entropy in the wiring, motor, and compressor that run the refrigerator. Shows a Carnot engine.

In the four segments the engine absorbs heat does the external work and finally absorbs its capacity for work while returning it to its initial state. It represents the most efficient heat engine.

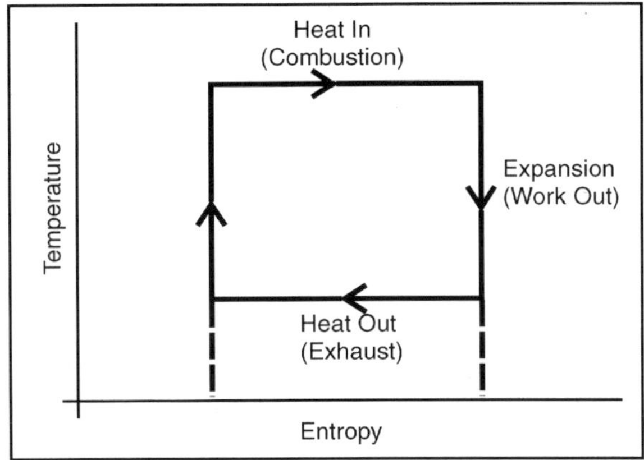

Consider a Second machine as shown in figure 4.16. Now we see that whether some cycle which is more efficient than the Carnot cycle exist. In other words, there exist a heat engine which is operating between temperatures T_2 and T_1 and extracts more work from the high temperature heat input Q_2 than $\in Q_2$. We see that $\in =1-T_1/T_2$ is the efficiency of the Carnot engine.

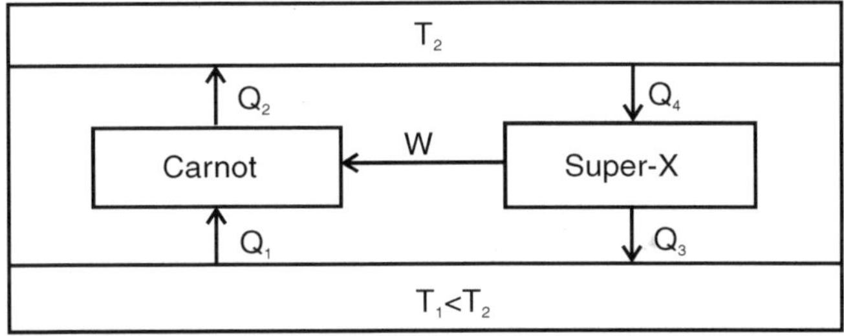

Consider a machine whose efficiency is greater than a Carnot engine as shown in figure.

This engine runs in reverse manner as a refrigerator, emitting heat Q_2 to the upper reservoir and absorbing Q_1 from the lower reservoir by using the work $W=Q_2-Q_1=\in Q_2$from the machine. The Super-X machine is operated in heat engine mode, emitting Q_3 to the lower reservoir and absorbing $Q_4= Q_3+W<W/e$ from the upper reservoir. The inequality indicates that the ratio of work produced and heat extracted from the upper reservoir, W/Q_4, is greater for the machine than for an equivalent Carnot engine.

Let us examine the net heat flow out of the upper reservoir, $Q_{upper}=Q_4-Q_2$. Since $Q_4<W/e=Q_2$, we find that $Q_{upper}<0$.

CARNOT ENGINE

A Carnot engine is used in a nuclear power plant. It receives 1500 MW by heat transfer with a source at 340C, and it rejects thermal waste to a nearby river at 27C. The river temperature rises by 3C which determines the:

- Efficiency of the power plant
- Power output of the power plant

The ideal Carnot engine was invented by French physicist Léonard which is theoretically perfect.

The diagram explains the working of the Carnot Engine.

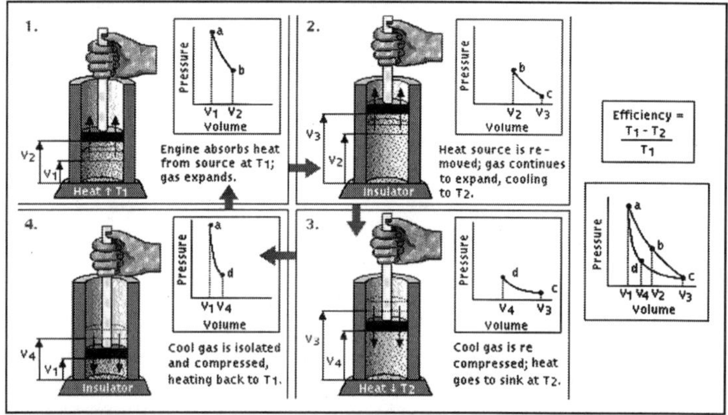

In this the efficiency of any engine depends on the difference between the highest and lowest temperatures. The greater the difference, the greater the efficiency.

THE THIRD LAW OF THERMODYNAMICS

The third law of thermodynamics is usually stated as a definition: the entropy of a perfect crystal of an element at the absolute zero of temperature is zero. At the absolute zero of temperature, there is zero thermal energy or heat. Since heat is a measure of average molecular motion, zero thermal energy means that the average atom does not move at all. Since no atom can have less than zero motion, the motion of every individual atom must be zero when the average molecular motion is zero. When none of the atoms which make up a perfectly ordered crystal move at all, there can be no disorder or different states possible for the crystal.

Taking the entropy of a perfect crystal of an element to be zero at the absolute zero of temperature establishes a method by which entropies of elements at any higher temperature can be determined. Since by definition dS = qrev/T, the mathematical integral of dS from zero to any higher temperature T is the integral, over that temperature range, of qrev/T. In other words. the difference S - S0 is the integral from zero to a temperature T of (Cp/T)dT. The molar heat capacity at any temperature is a measurable quantity, and so this difference can be determined experimentally.

The third law of thermodynamics is simply the statement that S0 is zero by definition for a pure element, and so if the heat capacity is measured under conditions of reversible heat flow, as it can be, and as a function of temperature at low temperatures, as it can be, then the entropy Sof a pure element at any temperature T is given by:

S = the integral from zero to T of (Cp/T)dT

The values we usually write as S0, the standard entropy of a substance, are actually the integral from zero to the standard temperature, 298.15 K, of (Cp/T)dT in order that the values of standard entropies of substances can be used with standard values for other thermodynamic functions such as enthalpy and free energy.

The value of the entropy of an element at any temperature, including 298.15 K which is the temperature of the standard entropy S0, can be obtained from careful measurements of the heat capacity of the element from the desired temperature down to absolute zero. Experimentally, chemists have been unable to reach the absolute zero of temperature, but measurements can be and have been made down to within 0.1 K of it and the heat capacity below measurable range can be accurately estimated. The entropy of the element is obtained from integration of the heat capacity measurements. If the crystal changes form, or melts, within the temperature range desired, then the entropy of that change must also be measured and included.

The entropies of compounds can be determined from thermodynamic measurements made on the reactions which form them from the elements or dissociate them to their constituent elements once the entropies of the elements themselves are established. The values of standard entropies of elements and compounds at 25°C of the thermodynamic properties of pure substances are all based upon the third law of thermodynamics.

ADIABATIC PROCESS

In thermodynamics, an adiabatic process or an isocaloric process is a thermodynamic process in which no heat is transferred to or from the working fluid. The term "adiabatic" literally means impassable (from Greek -äép-âáÖíåéí, not-through-to pass), corresponding here to an absence of heat transfer. Conversely, a process that involves heat transfer (addition or loss of heat to the surroundings) is generally called diabatic. Linguistically, diabatic is the opposite of adiabatic by the absence of the initial a.

For example, an adiabatic boundary is a boundary that is impermeable to heat transfer and the system is said to be adiabatically (or thermally) insulated; an insulated wall approximates an adiabatic boundary. Another example is the adiabatic flame temperature, which is the temperature that would be achieved by a flame in the absence of heat loss to the surroundings. An adiabatic process that is reversible is also called an isentropic process. Additionally, an adiabatic process that is irreversible and extracts no work is in an isenthalpic process, such as viscous drag, progressing towards a nonnegative change in entropy.

One opposite extreme—allowing heat transfer with the surroundings, causing the temperature to remain constant—is known as an isothermal process. Since temperature is thermodynamically conjugate to entropy, the isothermal process is conjugate to the adiabatic process for reversible transformations.

A transformation of a thermodynamic system can be considered adiabatic when it is quick enough that no significant heat is transferred between the system and the outside. At the opposite extreme, a transformation of a thermodynamic system can be considered isothermal if it is slow enough so that the system's temperature remains constant by heat exchange with the outside.

ADIABATIC HEATING AND COOLING

Adiabatic changes in temperature occur due to changes in pressure of a gas while not adding or subtracting any heat. Adiabatic heating occurs when the pressure of a gas is increased from work done on it by its surroundings, ie a piston. Diesel engines rely on adiabatic heating during their compression stroke to elevate the temperature sufficiently to ignite the fuel. Similarly jet engines rely upon adiabatic heating to create the correct compression of the air to enable fuel to be injected and ignition to then occur.

Adiabatic heating also occurs in the Earth's atmosphere when an air mass descends, for example, in a katabatic wind or Foehn wind flowing downhill.

Adiabatic cooling occurs when the pressure of a substance is decreased as it does work on its surroundings. Adiabatic cooling does not have to involve a fluid. One technique used to reach very low temperatures (thousandths and even millionths of a degree above absolute zero) is adiabatic demagnetisation, where the change in magnetic field on a magnetic material is used to provide adiabatic cooling. Adiabatic cooling also occurs in the Earth's atmosphere with orographic lifting and lee waves, and this can form pileus or lenticular clouds if the air is cooled below the dew point.

Rising magma also undergoes adiabatic cooling before eruption. Such temperature changes can be quantified using the ideal gas law, or the hydrostatic equation for atmospheric processes. It should be noted that no process is truly adiabatic. Many processes are close to adiabatic and can be easily approximated by using an adiabatic assumption, but there is always some heat loss. There is no such thing as a perfect insulator.

CONCEPT OF IRREVERSIBILITY

Irreversibility is a process which may be irreversible which cannot be directly reversed. This process is very common in mechanical machines. In the other words it is impossible to restore everything what is extracted in the system state. Consider a temperature, in a room. The amount of friction required to move a certain work in a friction state. Now from the energy conservation this work generate heat Q = W and increases the entropy as $\Delta S = \frac{W}{T} > 0$.

Hence, in a closed path, the heating due to friction, something that we can no longer reverse. In general, if we follow an irreversible process in completing a complete cycle, the entropy always increases.

That is

$$\Delta S \text{ closed path} > 0 : \text{Irreversible}$$

$$\Delta Q \text{closed path} > 0$$

Consider the formula

$$\Delta E = -\Delta W + \Delta Q$$

for entropy for reversible processes. The irreversible process starts from some initial state of the system and ends up in a final state. We construct an equivalent reversible process which also goes from the same initial to the same final state.

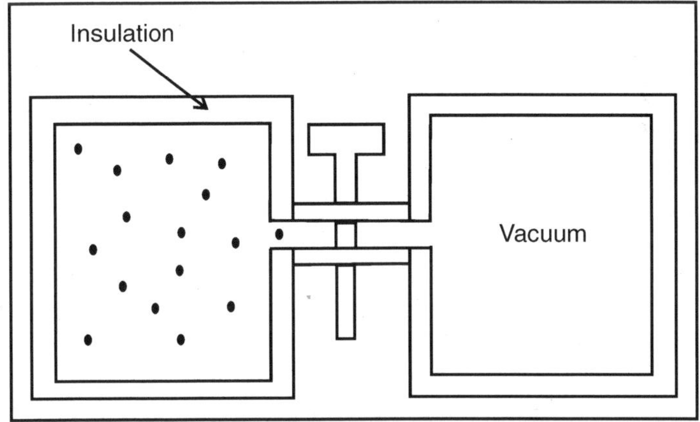

The gas expands into a vacuum and does no work or ΔW=0; furthermore, since it is thermally insulated there is no flow of heat, and hence ΔQ=0

7

Heat Engines

INTRODUCTION

Thermodynamics was invented, almost by accident, in 1825 by a young French engineer called Sadi Carnot who was investigating the theoretical limitations on the efficiency of steam engines. Although we are not particularly interested in steam engines, nowadays, it is still highly instructive to review some of Carnot's arguments. We know, by observation, that it is possible to do mechanical work upon a device , and then to extract an equivalent amount of heat , which goes to increase the internal energy of some heat reservoir. (Here, we use small letters and to denote intrinsically *positive* amounts of work and heat, respectively.) An example of this is Joule's classic experiment by which he verified the first law of thermodynamics: a paddle wheel is spun in a liquid by a falling weight, and the work done by the weight on the wheel is converted into heat, and absorbed by the liquid. Carnot's question was this: is it possible to reverse this process and build a device, called a *heat engine,* which extracts heat energy from a reservoir and converts it into useful macroscopic work? For instance, is it possible to extract heat from the ocean and use it to run an electric generator?

There are a few caveats to Carnot's question. First of all, the work should not be done at the expense of the heat engine itself, otherwise the conversion of heat into work could not continue indefinitely. We can ensure that this is the case if the heat engine performs some sort of cycle, by which it periodically returns to the same macrostate, but, in the meantime, has extracted heat from the reservoir and done an equivalent amount of useful work. Furthermore, a cyclic process seems reasonable because we know that both steam engines and internal combustion engines perform continuous cycles.

The second caveat is that the work done by the heat engine should be such as to change a single parameter of some external device (*e.g.,* by lifting a weight) without doing it at the expense of affecting the other degrees of freedom, or the entropy, of that device. For instance, if we are extracting heat from the ocean to generate electricity, we want to spin the shaft of the electrical generator without increasing the generator's entropy; *i.e.*, causing the generator to heat up or fall to bits. Let us examine the feasibility of a heat engine using the laws of thermodynamics. Suppose that a heat engine performs a single cycle. Since has returned to its initial macrostate, its internal energy is unchanged, and the first law of thermodynamics tell us that the work done by the engine must equal the heat extracted from the reservoir , so

The above condition is certainly a *necessary* condition for a feasible heat engine, but is it also a *sufficient* condition? In other words, does every device which satisfies this condition actually work? Let us think a little more carefully about what we are actually expecting a heat engine to do. We want to construct a device which will extract energy from a heat reservoir, where it is randomly distributed over very many degrees of freedom, and convert it into energy distributed over a single degree of freedom associated with some parameter of an external device.

Once we have expressed the problem in these terms, it is fairly obvious that what we are really asking for is a spontaneous transition from a probable to an improbable state, which we know is forbidden by the second law of thermodynamics. So, unfortunately, we cannot run an electric generator off heat extracted from the ocean, because it is like asking all of the molecules in the ocean, which are jiggling about every which way, to all suddenly jig in the same direction, so as to exert a force on some lever, say, which can then be converted into a torque on the generator shaft. We know from our investigation of statistical thermodynamics that such a process is possible, in principle, but is *fantastically improbable.*

The improbability of the scenario just outlined is summed up in the second law of thermodynamics. This says that the total entropy of an isolated system can never spontaneously decrease, so

$$\Delta S \geq 0$$

For the case of a heat engine, the isolated system consists of the engine, the reservoir from which it extracts heat, and the outside device upon which it does work. The engine itself returns periodically to the same state, so its entropy is clearly unchanged after each cycle. We have already specified that there is no change in the entropy of the external device upon which the work is done. On the other hand, the entropy change per cycle of the heat reservoir, which is at absolute temperature T_1, say, is given by

$$\Delta S_{reservoit} = \oint \frac{đQ}{T_1} = -\frac{q}{T_1}$$

where $đQ$ is the infinitesimal heat absorbed by the reservoir, and the integral is taken over a whole cycle of the heat engine. The integral can be converted into the expression because the amount of heat extracted by the engine is assumed to be too small to modify the temperature of the reservoir (this is the definition of a heat reservoir), so that T_1 is a constant during the cycle. The second law of thermodynamics clearly reduces to

$$\frac{-q}{T_1} \geq 0$$

or, making use of the first law of thermodynamics,

$$\frac{q}{T_1} = \frac{w}{T_1} \leq 0$$

Since we wish the work w done by the engine to be positive, the above relation clearly cannot be satisfied, which proves that an engine which converts heat directly into work is thermodynamically impossible.

A perpetual motion device, which continuously executes a cycle without extracting heat from, or doing work on, its surroundings, is just about possible according to Eq. In fact, such a device corresponds to the equality sign in Eq. which means that it must be completely *reversible*. In reality, there is no such thing as a completely reversible engine. All engines, even the most efficient, have frictional losses which make them, at least, slightly irreversible.

Thus, the equality sign in Eq. corresponds to an asymptotic limit which reality can closely approach, but never quite attain. It follows that a perpetual motion device is thermodynamically impossible. Nevertheless, the U.S. patent office receives about 100 patent applications a year regarding perpetual motion devices. The British patent office, being slightly less open-minded that its American counterpart, refuses to entertain such applications on the basis that perpetual motion devices are forbidden by the second law of thermodynamics.

According to Eq. there is no thermodynamic objection to a heat engine which runs backwards, and converts work directly into heat. This is not surprising, since we know that this is essentially what frictional forces do. Clearly, we have, here, another example of a natural process which is fundamentally irreversible according to the second law of thermodynamics. In fact, the statement

It is impossible to construct a perfect heat engine which converts heat directly into work is called Kelvin's formulation of the second law.

We have demonstrated that a *perfect heat engine*, which converts heat directly into work, is impossible. But, there must be some way of obtaining useful work from heat energy, otherwise steam engines would not operate. Well, the reason that our previous scheme did not work was that it decreased the entropy of a heat reservoir, at some temperature , by extracting an amount of heat per cycle, without any compensating increase in the entropy of anything else, so the second law of thermodynamics was violated.

How can we remedy this situation? We still want the heat engine itself to perform periodic cycles (so, by definition, its entropy cannot increase over a cycle), and we also do not want to increase the entropy of the external device upon which the work is done. Our only other option is to increase the entropy of some other body.

In Carnot's analysis, this other body is a second heat reservoir at temperature . We can increase the entropy of the second reservoir by dumping some of the heat we extracted from the first reservoir into it. Suppose that the heat per cycle we extract from the first reservoir is , and the heat per cycle we reject into the second reservoir is q_2 . Let the work done on the external device be w per cycle. The first law of thermodynamics tells us that

$$q_1 = w + q_2$$

Note that $q_2 < q_1$ if positive (*i.e.*, useful) work is done on the external device. The total entropy change per cycle is due to the heat extracted from the first reservoir and the heat dumped into the second, and has to be positive (or zero) according to the second law of thermodynamics. So,

$$\Delta S = \frac{-q_1}{T_1} + \frac{q_2}{T_2} \geq 0$$

We can combine the previous two equations to give

$$\frac{-q_1}{T_1} + \frac{q_1 - w}{T_2} \geq 0$$

or

$$\frac{w}{T_2} \leq q_1 \left(\frac{1}{T_2} - \frac{1}{T_1} \right)$$

It is clear that the engine is only going to perform useful work (*i.e.*, is only going to be positive) if $T_2 < T_1$. So, the second reservoir has to be *colder* than the first if the heat dumped into the former is to increase the entropy of the Universe more than the heat extracted from the latter decreases it. It is useful to define the efficiency of a heat engine. This is the ratio of the work done per cycle on the external device to the heat energy absorbed per cycle from the first reservoir. The efficiency of a perfect heat engine is unity, but we have already shown that such an engine is impossible. What is the efficiency of a realizable engine? It is clear from the previous equation that

$$\eta \equiv \frac{w}{q_1} \leq 1 - \frac{T_2}{T_1} = \frac{T_1 - T_2}{T_1}$$

Note that the efficiency is always less than unity. A real engine must always reject some energy into the second heat reservoir in order to satisfy the second law of thermodynamics, so less energy is available to do external work, and the efficiency of the engine is reduced. The equality sign in the above expression corresponds to a completely reversible heat engine (*i.e.*, one which is quasi-static). It is clear that real engines, which are always irreversible to some extent, are less efficient than reversible engines. Furthermore, all reversible engines which operate between the two temperatures T_1 and T_2 must have the *same* efficiency,

$$\eta = \frac{T_1 - T_2}{T_1}$$

irrespective of the way in which they operate.

Let us consider how we might construct one of these reversible heat engines. Suppose that we have some gas in a cylinder equipped with a frictionless piston. The gas is not necessarily a perfect gas. Suppose that we also have two heat reservoirs at temperatures and (where). These reservoirs might take the form of large water baths. Let us start off with the gas in thermal contact with the first reservoir. We now pull the piston out very slowly so that heat energy flows reversibly into the gas from the reservoir. Let us now thermally isolate the gas and slowly pull out the piston some more. During this adiabatic process the temperature of the gas falls (since there is no longer any heat flowing into it to compensate for the work it does on the piston). Let us continue this process until the temperature of the gas falls to . We now place the gas in thermal contact with the second reservoir and slowly push the piston in. During this isothermal process heat flows out of the gas into the reservoir. We next thermally isolate the gas a second time and slowly compress it some more. In this process the temperature of the gas increases.

We stop the compression when the temperature reaches . If we carry out each step properly we can return the gas to its initial state and then repeat the cycle *ad infinitum*. We now have a set of reversible processes by which a quantity of heat is extracted from the first reservoir and a quantity of heat is dumped into the second.

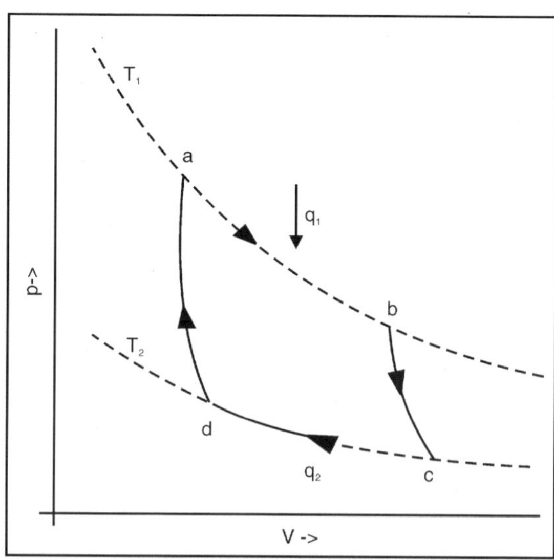

Fig. An Ideal Gas Carnot Engine

We can best evaluate the work done by the system during each cycle by plotting out the locus of the gas in a - diagram. The locus takes the form of a closed curve. The net work done per cycle is the "area" contained inside this curve, since $đw = pdV$ [if p is plotted vertically and horizontally, then pdV is clearly an element of area under the curve P(V)]. The engine we have just described is called a *Carnot engine*, and is the simplest conceivable device capable of converting heat energy into useful work.

For the specific case of an ideal gas, we can actually calculate the work done per cycle, and, thereby, verify Eq. Consider the isothermal expansion phase of the gas. For an ideal gas, the internal energy is a function of the temperature alone. The temperature does not change during isothermal expansion, so the internal energy remains constant, and the net heat absorbed by the gas must equal the work it does on the piston. Thus,

$$q_1 = \int_a^b pdV$$

where the expansion takes the gas from state a to state b. Since $pV = \nu RT$, for an ideal gas, we have

$$q_1 = \int_a^b \nu RT_1 \frac{dV}{V} = \nu RT_1 \text{In} \frac{V_b}{V_a}$$

Likewise, during the isothermal compression phase, in which the gas goes from state to state , the net heat rejected to the second reservoir is

$$q_2 = \nu RT_2 \text{In} \frac{V_c}{V_d}$$

Now, during adiabatic expansion or compression

$$TV^{\gamma - 1} = \text{constant}$$

It follows that during the adiabatic expansion phase, which takes the gas from state b to state c,

$$T_1 V_b^{\gamma-1} = T_2 V_c^{\gamma-1}$$

Likewise, during the adiabatic compression phase, which takes the gas from state d to state a,

$$T_1 V_a^{\gamma-1} = T_2 V_d^{\gamma-1}$$

If we take the ratio of the previous two equations we obtain

$$\frac{V_b}{V_a} = \frac{V_c}{V_d}$$

Hence, the work done by the engine, which we can calculate using the first law of thermodynamics,

$$w = q_1 - q_2$$

is

$$w = \nu R(T_1 - T_2) \text{In} \frac{V_b}{V_a}$$

Thus, the efficiency of the engine is

$$\eta = \frac{w}{q_1} = \frac{T_1 - T_2}{T_1}$$

which, not surprisingly, is exactly the same as in Eq.

The engine described above is very idealized. Of course, real engines are far more complicated than this. Nevertheless, the maximum efficiency of an ideal heat engine places severe constraints on real engines. Conventional power stations have many different "front ends" (*e.g.*, coal fired furnaces, oil fired furnaces, nuclear reactors), but their "back ends" are all very similar, and consist of a steam driven turbine connected to an electric generator.

The "front end" heats water extracted from a local river and turns it into steam, which is then used to drive the turbine, and, hence, to generate electricity. Finally, the steam is sent through a heat exchanger so that it can heat up the incoming river water, which means that the incoming water does not have to be heated so much by the "front end." At this stage, some heat is rejected to the environment, usually as clouds of steam escaping from the top of cooling towers. We can see that a power station possesses many of the same features as our idealized heat engine. There is a cycle which operates between two temperatures. The upper temperature is the temperature to which the steam is heated by the "front end," and the lower temperature is the temperature of the environment into which heat is rejected. Suppose that the steam is only heated to C (or K), and the temperature of the environment is C (or K). It follows from Eq. that the *maximum* possible efficiency of the steam cycle is

$$\eta = \frac{373 - 288}{373} \simeq 0.23$$

So, at least 77% of the heat energy generated by the "front end" goes straight up the cooling towers! Not be surprisingly, commercial power stations do not operate with C steam. The only way in which the thermodynamic efficiency of the steam cycle can be raised to an acceptable level is to use very hot steam (clearly, we cannot refrigerate the environment). Using C steam, which is not uncommon, the maximum efficiency becomes

$$\eta = \frac{673 - 288}{673} \simeq 0.57$$

which is more reasonable. In fact, the steam cycles of modern power stations are so well designed that they come surprisingly close to their maximum thermodynamic efficiencies.

HEAT

In physics, heat, symbolized by Q, is energy transferred from one body or system to another due to a difference in temperature. In thermodynamics, the quantity *TdS* is used as a representative measure of heat, which is the absolute temperature of an object multiplied by the differential quantity of a system's entropy measured at the boundary of the object. Heat can flow spontaneously from an object with a high temperature to an object with a lower temperature. The transfer of heat from one object to another object with an equal or higher temperature can happen only with the aid of a heat pump. High temperature bodies, which often result in high rates of heat transfer, can be created by chemical reactions (such as burning), nuclear reactions (such as fusion taking place inside the Sun), electromagnetic dissipation (as in electric stoves), or mechanical dissipation (such as friction). Heat can be transferred between objects by radiation, conduction and convection. Temperature is used as a measure of the internal energy or enthalpy, that is the level of elementary motion giving rise to heat transfer.

Heat can only be transferred between objects, or areas within an object, with different temperatures (as given by the zeroth law of thermodynamics), and then, in the absence of work, only in the direction of the colder body (as per the second law of thermodynamics). The temperature and phase of a substance subject to heat transfer are determined by latent heat and heat capacity. A related term is thermal energy, loosely defined as the energy of a body that increases with its temperature.

The first law of thermodynamics states that the energy of a closed system is conserved. Therefore, to change the energy of a system, energy must be transferred to or from the system. Heat and work are the only two mechanisms by which energy can be transferred to or from a control mass. Heat is the transfer of energy caused by the temperature difference. The unit for the amount of energy transferred by heat in the International System of Units SI is the joule (J), though the British Thermal Unit and the calorie are still occasionally used in the United States. The unit for the rate of heat transfer is the watt (W = J/s).

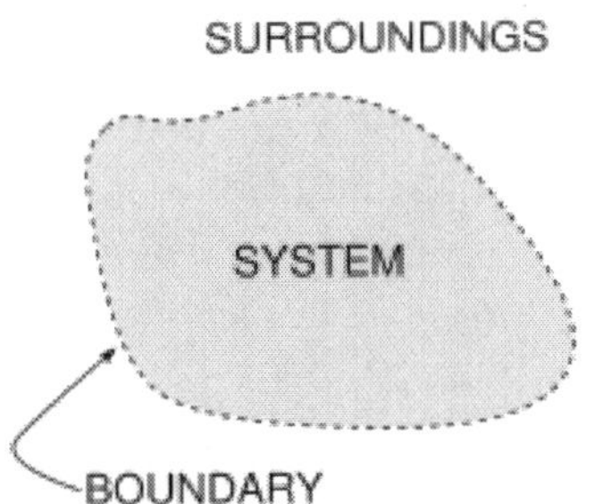

Heat Q can flow across the boundary of the system and thus change its internal energy *U*. Heat transfer is a path function (process quantity), as opposed to a point function (state quantity). Heat flows between systems that are not in thermal equilibrium with each other; it spontaneously flows from the areas of high temperature to areas of low temperature. When two bodies of different temperature come into thermal contact, they will exchange internal energy until their temperatures are equalized; that is, until they reach thermal equilibrium. The adjective *hot* is used as a relative term to compare the object's temperature to that of the surroundings (or that of the person using the term). The term heat is used to describe the flow of energy. In the absence of work interactions, the heat that is transferred to an object ends up getting stored in the object in the form of internal energy.

Specific heat is defined as the amount of energy that has to be transferred to or from one unit of mass or mole of a substance to change its temperature by one degree. Specific heat is a property, which means that it depends on the substance under consideration and its state as specified by its properties. Fuels, when burned, release much of the energy in the chemical bonds of their molecules. Upon changing from one phase to another, a pure substance releases or absorbs heat without its temperature changing. The amount of heat transfer during a phase change is known as latent heat and depends primarily on the substance and its state.

THERMAL ENERGY

Thermal energy is a term often confused with that of heat. Loosely speaking, when heat is added to a thermodynamic system its thermal energy increases and when heat is withdrawn its thermal energy decreases. In this point of view, objects that are hot are referred to as being in possession of a large amount of thermal energy, whereas cold objects possess little thermal energy. Thermal energy then is often mistakenly defined as being synonym for the word heat. This, however, is not the case: an object cannot possess heat, but only energy. The term "thermal energy" when used in conversation is often not used in a strictly correct sense, but is more likely to be only used as a descriptive word. In physics and thermodynamics, the words "heat", "internal energy", "work", "enthalpy" (heat content), "entropy", "external forces", etc., which can be defined exactly, i.e. without recourse to internal atomic motions and vibrations, tend to be preferred and used more often than the term "thermal energy", which is difficult to define.

NOTATION

The total amount of energy transferred through heat transfer is conventionally abbreviated as Q. The conventional sign convention is that when a body releases heat into its surroundings, $Q < 0$ (-); when a body absorbs heat from its surroundings, $Q > 0$ (+). Heat transfer rate, or heat flow per unit time, is denoted by:

$$\dot{Q} = \frac{dQ}{dt}.$$

It is measured in watts. Heat flux is defined as rate of heat transfer per unit cross-sectional area, and is denoted *q*, resulting in units of watts per square metre, though slightly different notation conventions can be used.

ENTROPY

In 1854, German physicist Rudolf Clausius defined the *second fundamental theorem* (the second law of thermodynamics) in the mechanical theory of heat (thermodynamics): "if two transformations which, without necessitating any other permanent change, can mutually replace one another, be called equivalent, then the generations of the quantity of heat Q from work at the temperature T, has the *equivalence-value*:"

$$\frac{Q}{T}$$

In 1865, he came to define this ratio as entropy symbolized by *S*, such that, for a closed, stationary system:

$$\Delta S = \frac{Q}{T}$$

and thus, by reduction, quantities of heat ä*Q* (an inexact differential) are defined as quantities of *TdS* (an exact differential):

$$\delta Q = TdS$$

In other words, the entropy function *S* facilitates the quantification and measurement of heat flow through a thermodynamic boundary.

DEFINITIONS

In modern terms, heat is concisely defined as energy in transit. Scottish physicist James Clerk Maxwell, in his 1871 classic *Theory of Heat*, was one of the first to enunciate a modern definition of "heat". In short, Maxwell outlined four stipulations on the definition of heat. One, it is "something which may be transferred from one body to another", as per the second law of thermodynamics. Two, it can be spoken of as a "measurable quantity", and thus treated mathematically like other measurable quantities. Three, it "can *not* be treated as a substance"; for it may be transformed into something which is not a substance, e.g. mechanical work. Lastly, it is "one of the forms of energy". Similar such modern, succinct definitions of heat are as follows:

- In a thermodynamic sense, heat is never regarded as being stored within a body. Like work, it exists only as *energy in transit* from one body to another; in thermodynamic terminology, between a system and its surroundings. When energy in the form of heat is added to a system, it is stored not as heat, but as kinetic and potential energy of the atoms and molecules making up the system.
- The noun heat is defined only during the process of energy transfer by conduction or radiation.
- Heat is defined as any spontaneous flow of energy from one object

to another, caused by a difference in temperature between two objects

- Heat may be defined as *energy in transit* from a high-temperature object to a lower-temperature object.
- Heat as an interaction between two closed systems without exchange of work is a pure heat interaction when the two systems, initially isolated and in a stable equilibrium, are placed in contact. The energy exchanged between the two systems is then called heat.
- Heat is a form of energy possessed by a substance by virtue of the vibrational movement, i.e. kinetic energy, of its molecules or atoms.
- Heat is the transfer of energy between substances of different temperatures.

HEAT CAPACITIES

Take-home message: Heat capacities are related to changes of entropy with temperature.

A heat capacity is the temperature change per unit heat absorbed by a system during a reversible process: . It is a poor name, since bodies don't contain heat, only energy, but we're stuck with it. (Note the difference between "heat capacity ()" and "specific heat capacity ()"; the latter is the heat capacity per kg or per mole - the units will make clear which.)

The heat capacity is is different for different processes. Useful heat capacities are those at constant volume or constant pressure (for a fluid). Since

$$đQ^{rev} = TdS$$

we have at constant volume $C_v dT = TdS$, so

$$C_v = T\left(\frac{\partial S}{\partial T}\right)_v$$

Similarly

$$C_p = T\left(\frac{\partial S}{\partial T}\right)_p$$

Furthermore at constant volume, no work is done on the system and so $đQ^{rev} = dE$; hence

$$C_V = \left(\frac{\partial E}{\partial T}\right)_V$$

Also (usefully for chemists),at constant pressure $đQ^{rev} = dH$ where H is the enthalpy, so

$$C_p = \left(\frac{\partial H}{\partial T}\right)_p$$

The specific heat capacity is the heat capacity per unit mass (or per mole). Heat capacities are not independent of temperature (or pressure) in general, but over a narrow temperature range they are often treated as such, especially for a solid.

Together with two of Maxwell's relations, we now have expressions for the partial derivatives of the entropy with respect to all easily manipulable variables (, ,). These can be used to derive expressions for the entropy change in real processes.

We can also derive a relation between , , and other measurable properties of a substance which can be checked experimentally: if is the isobaric thermal expansivity and k_T is the isothermal compressibility

$$\alpha = \frac{1}{V}\left(\frac{\partial V}{\partial T}\right)_P \text{ and } k_T = -\frac{1}{V}\left(\frac{\partial V}{\partial P}\right)_T$$

Then we have the relation

$$C_P - C_V = VT\frac{\alpha^2}{k_T}$$

which is always greater than zero. (The derivation is set as an exercise.) This relation is a firm prediction of thermal physics without any approximations whatsoever. It has to be true! For real gases and compressible liquids and solids it can be checked. For relatively incompressible liquids and solids it is hard to carry out processes at constant volume so C_V may not be well known and this equation can be used to predict it.

For one mole of a van der Waals gas this gives

$$C_P - C_V = R\left(1 - \frac{2a(V-b)^2}{RTV^3}\right)^{-1}$$

In the ideal gas limit $a, b \to 0$ this reduces to C_P-C_V=R as expected.

Entropy Changes

What is the entropy change during the expansion of a van der Waals gas for which C_V is a constant?

The equation of state for one mole of a van der Waals gas is

$$P = \frac{RT}{V-b} - \frac{a}{V^2};$$

represents the volume taken up by the finite size of the molecules, and is the reduction in pressure due to interactions between the molecules. In this way the two most important corrections neglected in the ideal gas are included. However the heat capacity at constant volume is still independent of temperature and volume, as in an ideal gas.

During an expansion, both the temperature and the volume may change. To calculate the change in entropy, we need and so note Many problems on this section of the course involve choosing variables, either or , writing in terms of infinitesimal changes in these variables as in the first line, and then using the definition of the heat capacities, and a Maxwell relation, to obtain something like the second line. Then we can calculate the total entropy change by integrating, first at constant and then at constant where in the second integration we used the fact that is constant. (We can do this integration because the two terms in are each functions of one variable only. Given that, the bottom line may be obvious to you without all the careful intermediate steps.)

Note that we have not said what kind of process (reversible or non-reversible, isothermal, adiabatic.) has taken place; that will go into the relation between and .

We can check our result against those we have already calculated for an ideal gas just by setting b=0 and using the correct C_V (*eg* 3R/2 for a monatomic gas), as follows: First, for an *isothermal* expansion ($T_1=T_2$) we get $\Delta S=R \ln(V_2/V_1)$ as before.

Also, for a *reversible adiabatic* expansion we can use $T_1 V_1^{\gamma-1} = T_2 V_2^{\gamma-1}$ so

$$\Delta S = R \ln \frac{V_2}{V_1} + C_V \ln \left(\frac{V_1}{V_2} \right)^{\gamma-1}$$

$$= \left(R - C_V \left(\frac{C_P}{C_V} - 1 \right) \right) \ln \frac{V_2}{V_1}$$

$$= 0$$

using $C_P - C_V = R$ for an ideal gas). But that is as expected: entropy changes for reversible, adiabatic processes are it always zero! conversely, we can use the fact that for a reversible adiabatic process to see that for a van der Waals gas, is constant. But so the exponent isn't

SYSTEMS WITH MORE THAN ONE COMPONENT

Take-home message: The chemical potential is important wherever the number of molecules in a system isn't fixed.

If we have a mixture of two substances present, the internal energy and all the other thermodynamical potentials will depend on how much of each is present, since there will be interactions between the molecules of each.

Here we focus on the Gibbs free energy, since the relevant conditions are usually those of fixed temperature and pressure.

We have where and are the number of molecules of each substance.This is easily generalised more than two components so now the first two partial derivatives are and as before. We *define* giving is called the chemical potential of substance but what is its significance.

First, imagine only one substance present. Then but is extensive, and and are intensive, so must be directly proportional to where is the Gibbs free energy per molecule, and so for a single component system, is just the Gibbs free energy per molecule. But for a two component system, depends not only on the extensive variable but also on the *ratio* , which is intensive.

The Gibbs free energy per molecule of one substance can depend on the concentration of the other. All we can say is is the *extra* Gibbs free energy per added molecule of substance 1. If substance 1 is ethanol and substance 2 water, the chemical potential of ethanol is different in beer (5%) and vodka (40%).

However for ideal gases, since there are no intermolecular interactions, the Gibbs free energies are independent and additive from and we see that the term is added to and also and an important use of chemical potentials is in chemical reactions. But it is important to physicists too. Not all reactions are chemical: in a neutron star, neutrons, protons and electrons can interact via ; they reach an equilibrium at which the chemical potentials on either side of the reaction are equal. This is heavily biased toward neutrons, because the chemical potential of the light electron is much higher for a given concentration than that of the heavy proton or neutron.

The chemical potential also governs systems which can exchange particles with a reservoir, and that is the context in which we will meet it in statistical physics.

Note: the use of to mean magnetic moment as well as chemical potential should never confuse, as magnets tend to have fixed numbers of atoms.

The last point will only become clear once we've done Gibbs distributions at the end of the course.

CHEMICAL REACTIONS

Take-home message: At equilibrium the chemical potential of reactants and products are equal. The treatment of chemical reactions is very like that of phase transitions. Again, we are considering conditions of constant temperature and pressure, and the question is the following: how far will a reaction go?

First consider the simplest case of a reaction with only one reactant and one product: . An example is the interconversion of n-pentane and isopentane (or pentane and methyl-butane, for those of us who learned our chemistry in the last thirty years).

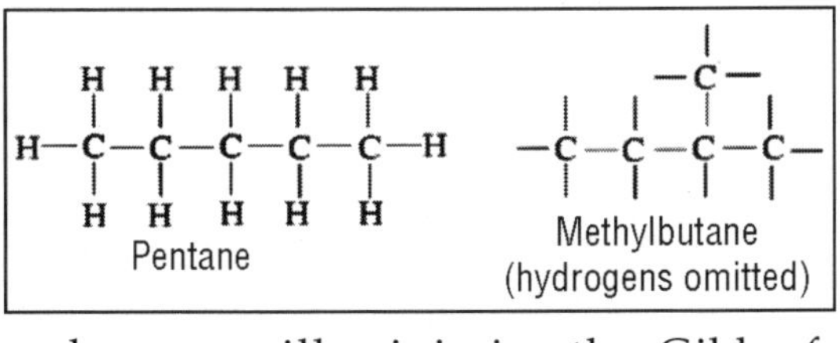

Pentane

Methylbutane (hydrogens omitted)

Spontaneous changes will minimise the Gibbs free energy With temperature and pressure fixed only the numbers of A and B can change. Since they can only interconvert, $dN_A = -dN_B$ and

$$dG = \mu_A dN_A + \mu_B dN_B = (\mu_A - \mu_B) dN_A$$

So if $\mu_A > \mu_B$, A will convert to B, but if $\mu_B > \mu_A$, the opposite will happen. So at equilibrium, when no further changes happen, the chemical potentials must be equal. (Remember that the chemical potentials are functions of concentration, so they will change as the reaction proceed.)

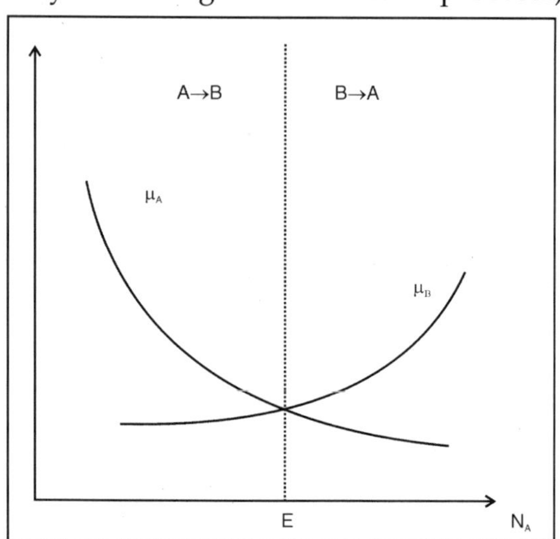

In the figure "E" marks the equilibrium concentration, at the point where $\mu A = \mu_B$

If there are more reactants or products, say $A + B \rightleftharpoons C + D$, the numbers of A, B, C and D change together $dN_A = dN_B = -dN_C = -dN_D$ So

$$dG = \mu_A dN_A + \mu_B dN_B + \mu_C dN_C + \mu_D dN_D$$
$$= (\mu_A + \mu_B - \mu_C - \mu_D) dN_A$$

and equilibrium is when

$$\mu_A + \mu_B = \mu_C + \mu_D$$

This result is general: equilibrium is reached when the sum of the chemical potential of the reactants equals that of the products.

COMBINATION OF FIRST AND SECOND LAW

We now wish to do some simple algebra to get the First and Second Laws Combined. Multiply Eq. I-2 by T_o to get

$$T_0 m_2 s_2 - T_0 m_1 s_1$$
$$= \left(\sum m_i T_0 S_i + \sum \frac{T_0}{T_i} Q_i\right) - \left(\sum m_e T_0 S_e + \sum \frac{T_0}{T_e} Q_e\right) + L_{cv}$$

Subtract Eq. I-3 from Eq. I-1 to get

$$m_2(u_2 - T_0 s_2) - m_1(u_1 - T_0 s_1)$$
$$= \left[\sum m_i (h_i - T_0 s_i) + \sum \left(1 - \frac{T_0}{T_i}\right) Q_i + \sum W_i\right]$$

$$-\left[\sum m_e(h_e - T_0 s_e) + \sum\left(1-\frac{T_0}{T_e}\right)Q_e + \sum W_e\right] - L_{cv}$$

Represent u - T_os by *a* and h - T_os by *b*. We now have our availability balance.

[Note in proof (9—9-97). We should refer to *a* as the Helmholtz availability *function* and *b* as the Gibbs availability *function,* two point functions, like energy, enthalpy, and entropy, that are thermodynamic properties of a simple homogeneous substance. These functions are employed in *lost work analysis,* the methodology employed here.

Exergy analysis is a competing or complementary methodology (depending upon one's viewpoint) that employs, instead, the thermodynamic property *exergy,* which, to make matters more confusing, is sometimes referred to as *availability.* (In the case of the 500 K hot water, used as an example in Chapter 2, the exergy is *equal* to the availability.) Exergy is essentially the difference between the availability function of the system and the availability function of the same atomic species when they are in mechanical, thermal, and chemical equilibrium with the surroundings. In this essay, we sometimes refer to the availability function as just the availability, likewise for the availability function balance, but we do not employ exergy analysis to such an extent that confusion could arise. When we use the term availability alone, we always mean the availability function

$$m_2 a_2 - m_1 a_1 = \left[\Sigma m_i b_i + \Sigma\left(1-\frac{T_0}{T_i}\right)Q_i + \Sigma W_i\right]$$

$$-\left[\Sigma m_e b_e + \Sigma\left(1-\frac{T_o}{T_e}\right)Q_e + \Sigma W_e\right] - L_{cv}$$

or, in rate form:

$$\dot{A} + \dot{L}_{cv} = \left[\Sigma f_i b_i + \Sigma\left(1-\frac{T_0}{T_i}\right)R_i + \Sigma P_i\right]$$

$$-\left[\Sigma f_e b_e + \Sigma\left(1-\frac{T_0}{T_e}\right)R_e + \Sigma P_e\right]$$

where A = m<a>, mass times average availability, i.e., availability per kilogram. When we wish to denote rate of accumulation of availability, say, per unit time, we merely place a dot over the symbol for availability. This is standard practice among physicists and engineers and applies to anything; i.e., if X stand for volume of beer drunk, (spoken and sometimes written 'X dot') stands for the volume of beer drunk per unit time at a particular time of interest or over a period of time such that the rate of guzzling remained constant. Although averages are denoted normally by angular brackets, viz., we may omit the brackets when no confusion can arise, in which case X dot stands for the average rate of guzzling during the period of interest. (Aren't you glad you decided to read this?) But, we haven't said what kind of availability function we are talking about and, in keeping with Murphy's 352d Law, there are two kinds (represented by *a* and *b*).

Amazingly, despite the incredible importance of the quantities *a* and *b*, they do not have decent names even. Perhaps, this is indicative of a less than felicitous point of view adopted by scientists and engineers over the years. To assist our memories, let us call $a = u - T_0 s$ the Helmholtz availability function (since u - Ts is the well-known Helmholtz function) and $b = h - T_0 s$ the Gibbs availability function (since h - Ts is the well-known Gibbs function).

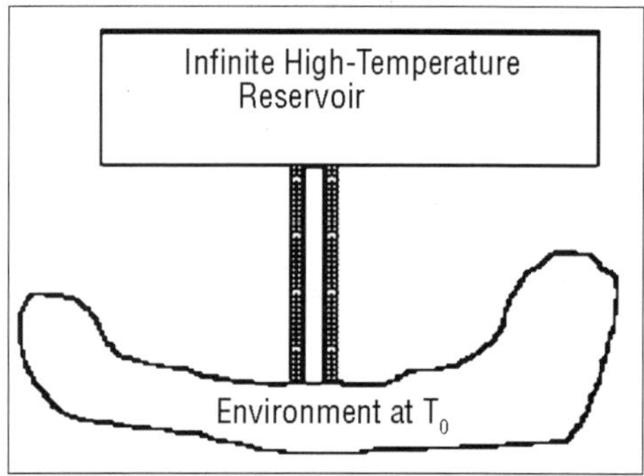

Fig. Diagram to Illustrate Lost Work

In rate equations, we find it confusing to employ derivative notation. If m_{cs} is the quantity of mass crossing the control surface, we shall refer to the rate at which mass crosses the control surface as f_{cs} (for flow). Similarly, R_{cs} is the rate of heat transfer across the control surface, and P_{cs} (for power) is the rate at which work is done on or by the control volume. These have convenient mathematical equivalents, which you may know already or will learn later.

We may employ Eq. I-6 to familiarize ourselves with important thermodynamic concepts. In particular, let us consider a closed system in steady state. The accumulation term, , is zero and the entrance and exit terms, år$_i$b$_i$ and år$_e$b$_e$, are both zero. Suppose, in addition, that no work is done on or by the control volume. Eq. I-6 is reduced to

$$\dot{L}_{cv} = \Sigma\left(1 - \frac{T_0}{T_i}\right)R_i - \Sigma\left(1 - \frac{T_0}{T_e}\right)R_e$$

Let us select a concrete example to see how this equation makes clear the meaning of lost work. Suppose we have a long metal rod – well-insulated except for the ends – touching a practically infinite high-temperature source like a large boiler at temperature T_H at one end (of the rod) and the atmosphere or ground at temperature T_0 at the other as shown in Fig. I-3. The control surface is taken to be the outside of the insulation and the bare metallic ends of the rod.

The heat influx rate is $R_i = R_H$, whilst the heat efferent rate is $R_e = R_L = R_o$. The insulation is important because, under these conditions, the heat out term $R_e$will be multiplied by $1 - T_o/T_o = 0$, which would not be the case if heat leaked out the sides at higher temperatures. To maintain steady state, we must have a positive heat term, $R_H = R_i$, entering the control volume from the boiler at temperature T_H. Then the lost work is easily seen to be precisely the work that would have been done by a reversible (Carnot) engine operating between a heat source at temperature T_H and rejecting heat to a heat sink at temperature T_o

$$\dot{L}_{cv} = \left(1 - \frac{T_0}{T_H}\right) R_H$$

[This term 1- T_o/T_X occurs so often that we find it expedient to further simplify our equations by denoting it C_X (for Carnot). The above equation could have been written

$$\dot{L}_{cv} = C_H R_H$$

which is perhaps going too far.] Thus, L really does represent the work we could have gotten from an ideal process but didn't get from our real process, which wasted the high-temperature heat that was added to it. Question: Where was the irreversibility in this system?

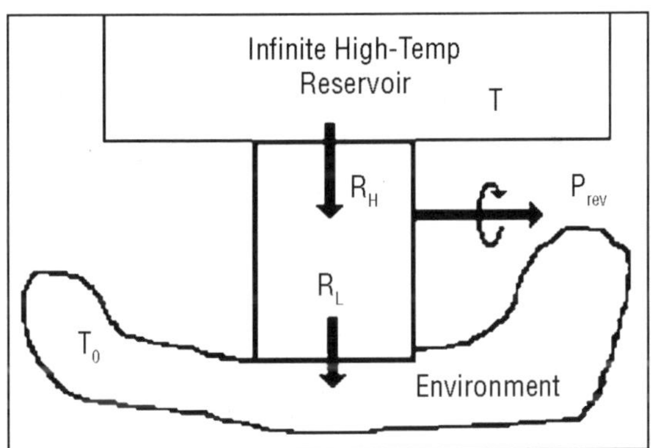

Fig. A Completely Reversible Device

In a *reversible steady-state* process conducted upon a *closed system*, a heat engine, say, that produces work, heat enters at temperature and leaves at temperature T_o, in which case Eq. I-6 reduces to

$$P_e = \left(1 - \frac{T_0}{T_i}\right) R_i - \left(1 - \frac{T_0}{T_0}\right) R_e$$

Since the accumulation term, A dot, equals zero (steady state), the terms representing mass entering and leaving are zero (closed system), and the lost work term, L_{cv} dot, equals zero (reversibility). The work done by the control volume is equal to the*reversible work*, i.e., the maximum amount of work that can be extracted from R_i at temperature T_i. Thus,

$$P_{rev} = \left(1 - \frac{T_0}{T_i}\right) R_i - \left(1 - \frac{T_0}{T_0}\right) R_e$$

as shown in Fig. I-4. But, the second term in parentheses is identically zero, therefore the equation reduces to

$$P_e = \left(1 - \frac{T_0}{T_i}\right) R_i = C_i R_i$$

Thus, the control volume is a heat engine with an efficiency $h = W_e / Q_i = 1 - T_o / T_i = C_i$. This is precisely the efficiency of a Carnot engine. We know that no device can have an efficiency as high as that of a Carnot engine except a Carnot engine itself; therefore, our control volume must *be* a Carnot engine, the imaginary device, discussed above, whose efficiency can be approached but never attained.

Finally, let us consider a reversible steady-state process with one stream entering and one stream leaving. We wish to know the maximum amount of work that could be obtained from such a process. This serves as an upper bound on the work that we can expect to obtain from a process with this input and this output.

$P_{rev} = f_i b_i - f_e b_e$.

HEAT ENGINES AND REFRIGERATORS

In any heat engine, heat is extracted from a hot source (eg hot combustion products in a car engine). The engine does work on its surroundings and waste heat is rejected to a cool reservoir (such as the outside air). It is an experimental fact that the waste heat cannot be eliminated, however desirable that might be. Indeed in practical engines, more of the energy extracted from the hot source is wasted than is converted into work.

The efficiency η of a heat engine is the ratio of the work done to the heat extracted from the hot source Q_H:

$$\eta_{engine} = \frac{W}{Q_H}$$

If Q_C is the rejected heat, we have by conservation of energy

$$Q_H - Q_C = W$$

An example of an efficiency calculation for a particular cycle (the Otto cycle) can be found here.

Heat engines can also be used to pump heat from colder to hotter bodies. This always needs an input of work. Examples are fridges, air conditioners and heat pumps. These are all essentially the same, but the in the first two the purpose is to cool (or keep cool) a fridge, room or building and in the last, the purpose is to heat (or keep hot) a building. In these cases what we call the efficiency is different. In general

$$\text{efficiency} = \frac{\text{desired output}}{\text{costly input}}$$

In the heat engine, it is Q_H that is costly (eg, petrol!) and that we want out, so as already given,

$$\eta_{\text{engine}} = \frac{W}{Q_H} < 1 \quad \text{always}$$

In the fridge or air conditioner, it is the work that is costly (electricity) and the desired result is the extraction of Q_C from the room or fridge, so

$$\eta_{\text{fridge}} = \frac{Q_C}{W} > 1 \quad \text{usually}$$

In the heat pump, it is again the work that is costly (electricity) and the desired result is the addition of Q_H to the building, so

$$\eta_{\text{pump}} = \frac{Q_H}{W} > 1 \quad \text{always}$$

These are sometimes called "coefficients of thermal performance" because students are worried by "efficiencies" greater than one. There would be no point in the heat pump if its efficiency, as we've defined it, were less than one—we'd just use an electric heater (efficiency exactly one) instead!

Note that real engines are optimized to perform forward or backwards, and are not reversible in the technical sense (there is friction, processes are not quasistatic). Thus relative sizes of W, Q_H and Q_C will depend on the direction. However for idealized, reversible engines only the signs will change, and we have

$$\eta_{\text{pump}}^{\text{rev}} = \frac{1}{\eta_{\text{engine}}^{\text{rev}}}$$

Often we are concerned with engines operating between only two reservoirs - one hot and one cold. In that case there is a standard way of denoting a heat engine, given below with the corresponding cycle in the P – V plot. If all the arrows are reversed it represents a heat pump or fridge.

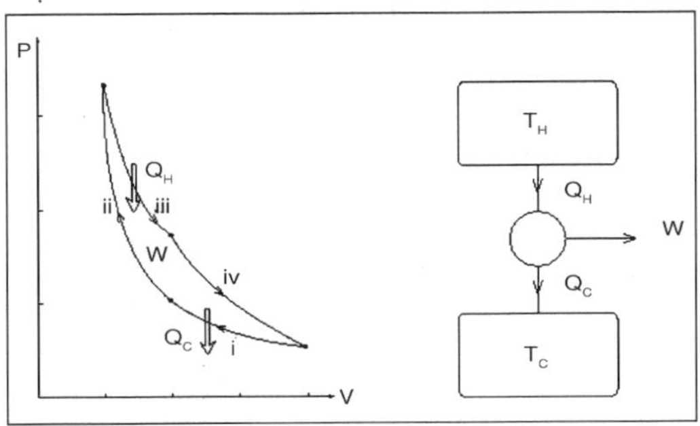

Questions about heat engines and pumps are often concerned with the most efficient possible engines, not with specific cycles.

THE OTTO CYCLE

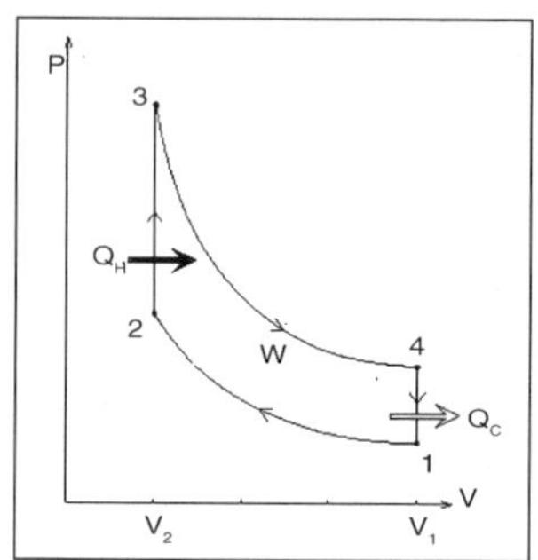

Here is a particular example of a heat engine - the Otto cycle, which approximates the working cycle of a car engine. The work done is the area enclosed by the curve in the $P - V$ plane. The expansion and compression stokes are adiabatic, so heat enters and leaves only during the constant volume phases.

A pretty animation (which is more realistic than above, as it includes the exhaust stroke) can be found.

The efficiency of a reversible Otto cycle for an ideal gas is

$$\eta = 1 - \left(\frac{V_2}{V_1}\right)^{\gamma - 1}$$

SECOND LAW OF THERMODYNAMICS

There are two classic statements of the second law of thermodynamics One due to Kelvin and Planck:

It is impossible to construct an engine which, operating in a cycle, will produce no other effect than the extraction of heat from a reservoir and the performance of an equivalent amount of work.

And another due to Clausius: It is impossible to construct an refrigerator which, operating in a cycle, will produce no other effect than the transfer of heat from a cooler body to a hotter one.

Note the careful wording: of course it is possible to think of processes which convert heat into work (expansion of a hot gas in a piston) or which pass heat from a cool to a hot body (real fridges) but other things change as well (the gas ends up cooler; you have an electricity bill to pay). The bit about "operating in a cycle" ensures that the engine is unchanged by the process.

EQUIVALENCE OF KELVIN AND CLAUSIUS STATEMENTS

The two statements of the second law may not appear to have anything to do with one another, but in fact each one implies the other.

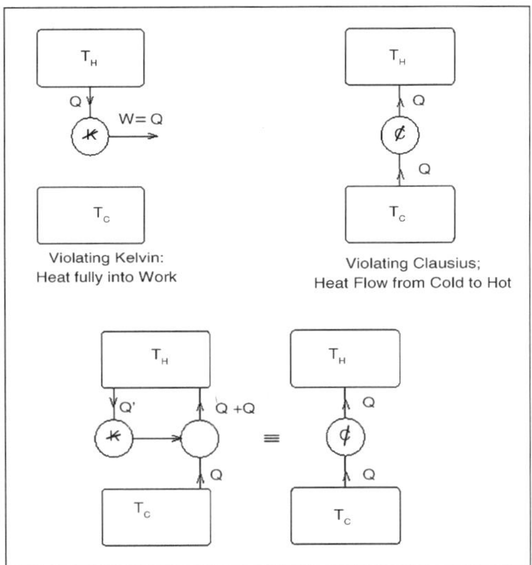

The top line shows yellow engines representing hypothetical engines violating each of the two statements of the second law. The second line shows that an engine which violates Kelvin's statement, together with a normal heat pump, violates Clausius's statement. A similar drawing can be used to show that an pump which violates Clausius's statement, together with a normal heat engine, violates Kelvin's statement. Hence the two statements—which seem quite different—are actually equivalent.

CARNOT ENGINE

A Carnot engine is, simply, a reversible engine acting between only two heat reservoirs. That means that all processes are either isothermal (heat transfer at a constant temperature) or adiabatic (no heat transfer). By contrast the Otto cycle, which has heating at constant volume, would need a whole series of heat reservoirs at incrementally higher temperatures to carry out the heating reversibly.

Carnot's theorem says that a reversible engine is the most efficient engine which can operate between two reservoirs. If you want to see the proof, see here. An equally important corollory is that any reversible engine working between two heat reservoirs has the same efficiency as any other, irrespective of the details of the engine.

Much is made of the fact that the Carnot engine is the most efficient engine. Actually, this is not mysterious. First, if we specify only two reservoirs, then all it says is that a reversible engine is more efficient than an irreversible engine, which isn't surprising (no friction...) Second, we will see that the efficiency of a Carnot engine increases with the temperature difference between the reservoirs. So it makes sense to use only the hottest and coldest heat baths you have available, rather than a whole series of them at intermediate temperatures. But the independence of the details of the engine is rather deeper, and has far-reaching consequences.

As a result, if we can calculate the efficiency for one Carnot engine, we know it for all. We can calculate it for an ideal gas Carnot cycle, and find:

$$\eta_{\text{carnot}} = 1 - \frac{T_C}{T_H},$$

Hence this is true for *all* Carnot engines.

By comparing with the definition of the efficiency of any heat engine, $\eta = W / Q_H = 1 - Q_C / Q_H$, we get the even more useful relation:

$$\frac{Q_C}{Q_H} = \frac{T_C}{T_H}.$$

Carnot developed the concept of reversibility and showed that no engine could be more efficient than a reversible one before either the first or second law of thermodynamics had been formulated! See here for a link to his essay, and to a good description of his contribution to thermodynamics.

Carnot Wins

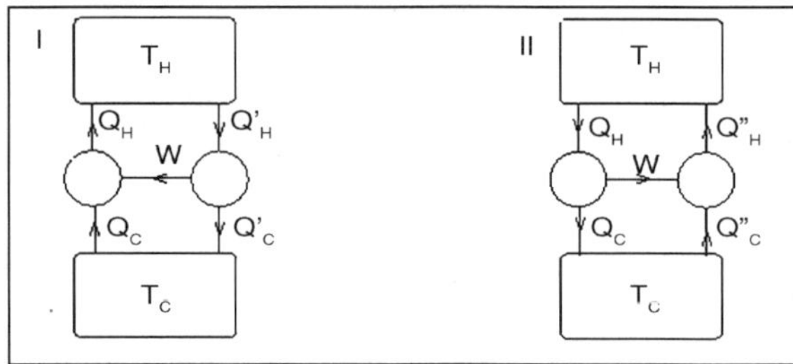

In the above the green engines/pumps are Carnot, and the brown ones are irreversible. Because the Carnot engine is reversible, if the engine efficiency is η_C, the pump efficiency is $1/\eta_C$ In the first case, a Carnot pump is driven by an engine. Overall, to avoid violating Clausius's statement of the second law, there must be no net heat transfer to the hot reservoir, so $Q_{H'} \geq Q_H$. Also, by conservation of energy, $Q_H - Q_C = Q'_H - Q'_C = W$. Thus

$$\frac{W}{Q'_H} = \frac{Q_H}{Q'_H}\frac{W}{Q_H}$$

So $\qquad \eta_{\text{engine}} \leq \eta_{\text{carnot}}$

In the second case, a Carnot engine drives a pump. Now we need $Q_H \geq Q_{H''}$. Thus

$$\frac{Q''_H}{W} = \frac{Q''_H}{Q_H}\frac{Q_H}{W}$$

$$\text{So} \qquad \eta_{\text{pump}} \leq \frac{1}{\eta_{\text{carnot}}}$$

So no engine can be more efficient than a Carnot engine, and no pump can be more efficient than a Carnot pump.

If the brown pumps were in fact reversible, there could be no overall heat flow, since heat flow from a hot to a cold body is an irreversible process. In that case the inequalities would become equalities.

Thus any reversible engine working between two heat reservoirs has the same efficiency as any other.

Efficiency of ideal gas Carnot cycle

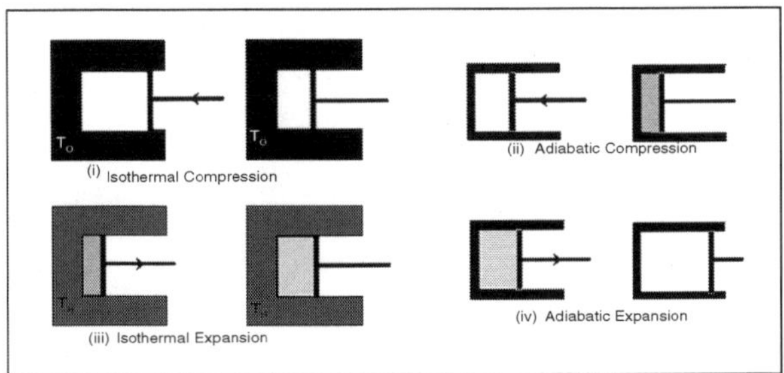

Above we see the four steps which make up an ideal gas Carnot cycle, two isothermal, (i) and (iii), and two adiabatic, (ii) and (iv). Below the cycle is sketched on a *P* – *V* plot.

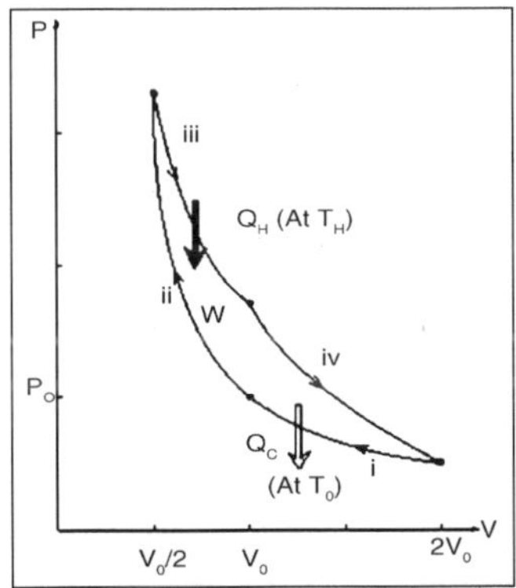

We have already analysed this particular cycle We worked out the work for the four steps. We note further that the heat exchanged during the isothermal steps is just minus the work done, since the internal energy is unchanged.

Thus we have $W = nR\ln 2(T_H - T_0)$ and (from stage iii) $Q_H = nRT_H \ln 2$, so

$$\eta = \frac{W}{Q_H} = 1 - \frac{T_0}{T_H}.$$

Note this cycle wasn't general, because we chose to make both the isothermal and adiabatic stages volume-halving. But form of the answer is general.

REVERSIBLE PROCESSES

Imagine a cylinder, with a perfectly smooth piston, which contains gas. If you push with a force only just large enough to overcome the internal pressure, the volume will start to decrease slowly. Then if you decrease the force only slightly, the volume will start to increase. This is the hallmark of a reversible process: an infinitesimal change in the external conditions reverses the direction of the change. Heat flow is only reversible if the temperature difference between the bodies is infinitesimally small.

Reversible processes require the absence of friction or other hysteresis effects. They must also be carried out infinitesimally slowly. Otherwise pressure waves and finite temperature gradients will be set up in the system, and irreversible dissipation and heat flow will occur.

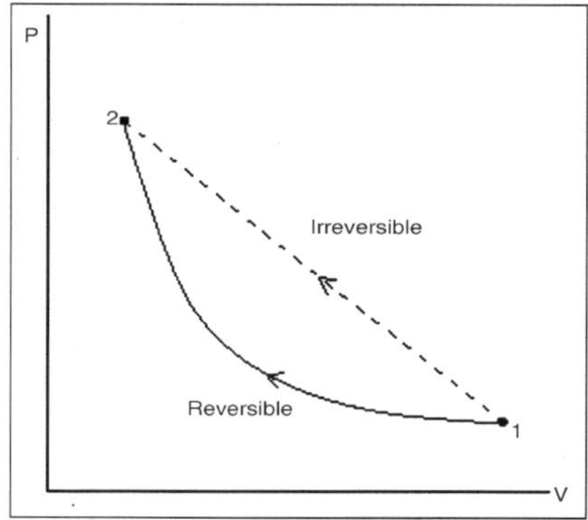

Because reversible processes are very slow, the system is always very nearly in equilibrium at all times. In that case all its state variables are well defined and uniform, and the state of the system can be represented on a plot of, for instance, pressure versus volume. A finite reversible process passes through an infinite set of such states, and so can be drawn as a solid line on the plot. During an irreversible process the system is not in an equilibrium state, and so cannot be represented on the plot; an irreversible process is often drawn as a straight dotted line joining the initial and final equilibrium states. The work done during the process is *not* then equal to the area under the line, but will be greater than that for the corresponding reversible process.

EXAMPLES

The master-equations for generic heat-engine problems are the conservation of energy and the heat-temperature relation for a Carnot engine:

$$W = Q_H - Q_C \qquad \text{and} \qquad \frac{Q_C}{Q_H} = \frac{T_C}{T_H}$$

The question will give you one of, and Q_H, Q_C and W the master-equations give you the other two.

Example: A power station contains a heat engine operating between two heat reservoirs, one consisting of steam at C and the other consisting of water at C. What is the maximum amount of electrical energy which can be produced for every Joule of heat extracted from the steam?

Answer: A power station contains a heat engine operating between two heat reservoirs, one consisting of steam at C and the other consisting of water at 100°C. What is the maximum amount of electrical energy which can be produced for every Joule of heat extracted from the steam?

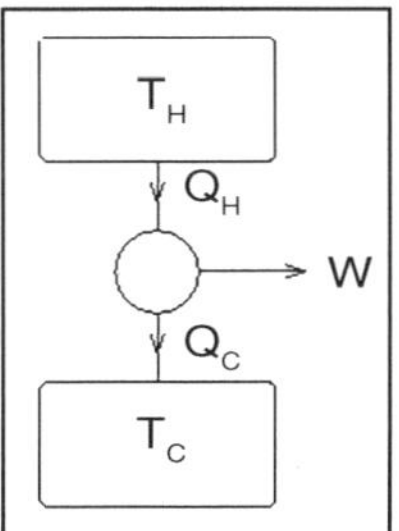

The maximum efficiency comes from using a Carnot (reversible) engine, for which $Q_H / Q_C = T_H / T_C$, and so

$$W = Q_H - Q_C = Q_H\left(1 - \frac{T_C}{T_H}\right)$$

With T_H = 373 K and T_C = 293 K, for every Joule of heat extracted from the hot reservoir (Q_H = 1 J), W_{max} = 0.21 J.

Example: When a fridge stands in a room at C, the motor has to extract 500W of heat from the cabinet, at C, to compensate for less than perfect insulation. How much power must be supplied to the motor if its efficiency is 80% of the maximium achievable?

Answer: When a fridge stands in a room at 20°C, the motor has to extract 500W of heat from the cabinet, at 4°C, to compensate for less than perfect insulation. How much power must be supplied to the motor if its efficiency is 80% of the maximium achievable?

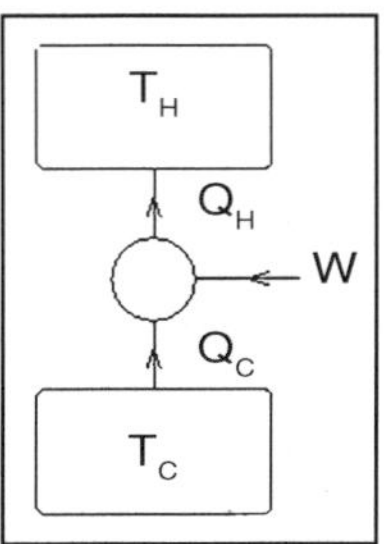

The maximum efficiency comes from using a Carnot (reversible) fridge, for which $Q_H / Q_C = T_H / T_C$. Here Q_C is known, and so

$$W = Q_H - Q_C = Q_C\left(\frac{T_H}{T_C} - 1\right)$$

(In this question W, Q_H and Q_C will refer to energy transfer per second, measured in Watts.) With $T_H = 293$K and $T_C = 277$ K, $Q_C = 500$W gives $W_{max} =$ 29W. However the real fridge works at 80% of the maximum efficiency, so $W = (29 / 0.8) = 36$ W.

PRACTICAL ENGINES

- A number of practical heat engines have been invented and are now in use. One of the most elegant, though one of the least-used, is the "Stirling cycle" engine. This is an "external combustion" engine, meaning that it is powered by an external source of heat. Any reasonably practical heat source can be used, such combustion of gasoline, kerosene, wood or manure, or sunlight focused by a mirror. All the Stirling cycle engine needs to run is a source of heat on one side and a cooling sink on the other.

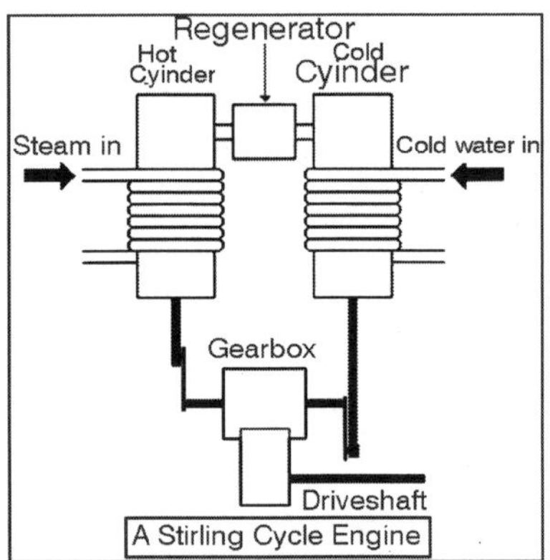

A Stirling Cycle Engine

The Stirling cycle engine's indifference to fuel source is its main advantage, and it is of some use in undeveloped countries where access to fuels is limited. Its disadvantage is that it has a poor "power to weight ratio (PWR)". All other engines in common use can provide the same power for much less weight. The illustration below is an example of a simple (and idealized) implementation of a St ove.

The gearbox steps the engine through a four-part cycle as follows:

- While the piston in the cold cylinde r is stopped at the top of its travel, heat input into the hot cylinder causes the piston there to move downward. This is the power stroke, an isothermal expansion process.

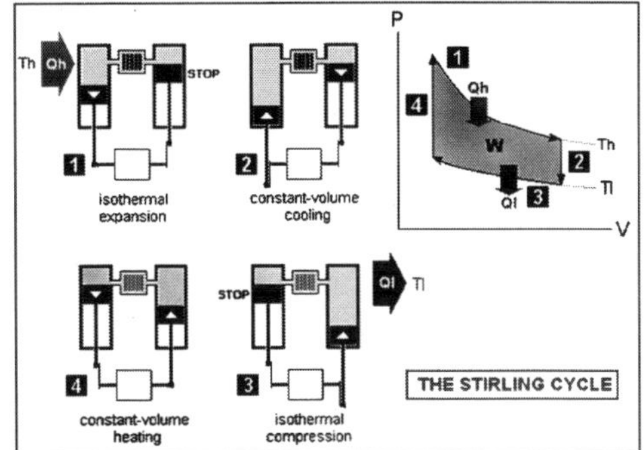

- The piston in the hot cylinder is then moved to the top of its travel while the piston in the cold cylinder is moved to the bottom of its travel. The volume of the working gas in the engine remains the same, while heat from the gas is stored in the regenerator. This is a constant-volume cooling process.
- The piston in the hot cylinder is stopped at the top of its travel while the piston in the cold cylinder is moved up the cylinder to half its full travel. This compresses the working gas, but the temperature remains constant because heat is being drawn off by the cold water. This is an isothermal compression process.
- Now the piston in the hot cylinder is moved downward to half its full travel, while the piston in the cold cylinder is moved to the top of its travel. This is a constant-volume process, in which heat from the regenerator is fed back into the gas flowing into the hot cylinder. The cycle is now ready to begin all over again.

This two-cylinder configuration is sometimes known as an "alpha" Stirling engine. In practice, Stirling engines are often built in a "beta" or "displacer" configuration, in which both pistons are in one cylinder, on top of one another and operated by concentric rods, and the regenerator runs up the sides of the cylinder, with ports between the two pistons and above the top piston.

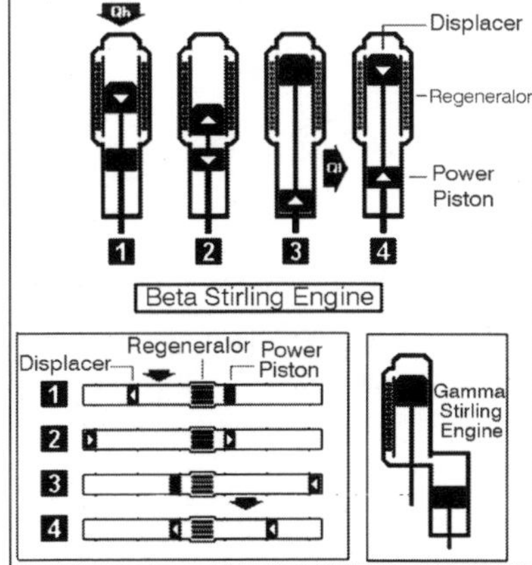

The top piston is called the "displacer", since its function is to "displace" the working fluid, while the bottom piston is called the "power piston", since it provides the drive power out of the engine.

The beta Stirling engine's operating principles are exactly the same as those of an alpha Stirling engine, though it is trickier to understand. There is also a "gamma" Stirling engine configuration, which is exactly the same as the beta configuration except that it is rearranged a bit to avoid the use of concentric rods.

The illustration below shows a beta Stirling engine stepping through its cycles, with an idealized Stirling engine shown below to help clarify its operation, along with a gamma configuration engine.

In practice, the gearing system that drives a beta Stirling engine is designed so that a piston may slow down but won't ever actually stop, except for the instant between reversals of direction. This construction "rounds off" the edges of the PV diagram, but otherwise the operation is the same.

- The most popular heat engine in service is the automotive internal combustion engine, known formally as the "Otto cycle" engine and more popularly as the "four-stroke" engine.

Operation of the Otto cycle is conceptually simple:

- In the first part of the cycle, the "intake stroke", the intake valve opens and the piston moves down to its lowest position. This draws in a gasoline-air mixture at (constant) atmospheric pressure.

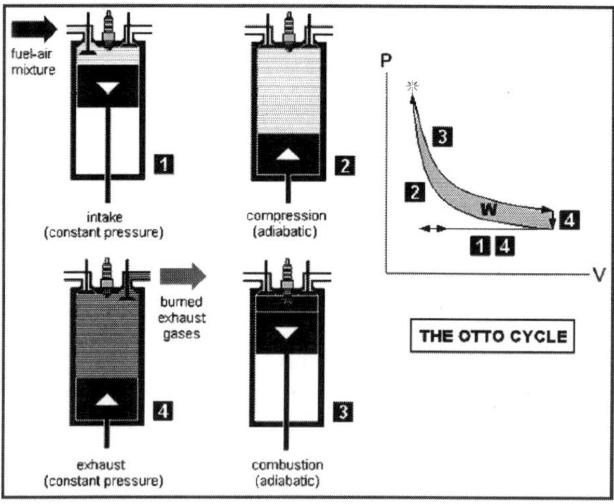

- In the "compression stroke", the intake valve is closed and the piston moves to its top position, compressing the fuel-air mixture. This is done quickly enough so that this part of the cycle is effectively adiabatic.
- In the "power stroke", an electric spark plug ignites the compressed fuel-air mixture, driving the piston to its lowest position.

- In the "exhaust stroke", the exhaust valve opens and the piston moves up, exhausting the combustion products of the fuel-air mixture at (constant) atmospheric pressure.

Analysis of the Otto cycle engine shows that its efficiency is mostly dependent on the "compression ratio", that is, the ratio of maximum compression of the fuel-air mixture to atmospheric pressure. The greater the compression ratio, the more efficient the engine.

However, the Otto cycle engine is limited on the level of compression it can obtain, since at high compressions the temperatures and pressures will cause the fuel-air mixture to ignite spontaneously before the piston reaches the top of its travel. This phenomenon is known as "engine knock".

The "Diesel cycle" engine avoids this problem. It has the same general four-stroke operational cycle as the Otto cycle engine, but uses a fuel injection system to spurt fuel directly into the cylinder at the end of the compression stroke, permitting higher compression ratios. The Diesel engine uses heavier fuels than the Otto cycle engine. It does not use spark plugs, instead using "glow plugs" that are heated and assist ignition of the fuel-air mixture when the compression reaches the proper level.

A second variation on the Otto cycle is the "two-stroke" engine, in which the intake and expansion cycles are combined, as are the compression and exhaust cycles. This means that the two-stroke engine has, in the limit, twice as much power for a given RPM than a four-stroke engine and is correspondingly lighter.

The problem with two-stroke engines is that they unsurprisingly burn very dirty, and so have been generally banned by air-pollution regulations. However, experimental two-stroke engines have been built that use electronic control systems to meet air-pollution regulations.

- There is a very wide range of configurations of four-stroke and two-stroke engines air or water cooled; or pistons in flat, inline, vee, or radial configurations. There are also exotic variations on the theme, such as the well-known "Wankel" rotary engine, which instead of a piston in a cylinder uses a rotor in the form of a "fat" triangle rotating inside a combustion chamber with a cross section like that of a kidney bean.

  Describing all the details and variations would be a major document in itself; it should be enough to say here that no matter what the configuration is, all such engines work on the same basic principles.

One configuration, the "free piston engine", is so unusual that it is worth adding here. The name is due to the fact that this class of engine doesn't have any driveshaft: the pistons just pop back and forth and down without being connected to anything, being driven forward by combustion to compress a closed-off air space and being driven back again. This sounds pointless, but the motion of the pistons can be used directly to pump gases or liquids. The diagram below shows a free piston engine used as a water pump. Its basic operation uses the four-stroke Diesel cycle; there is no reason in principle that it couldn't use a four-stroke Otto cycle or a two-stroke cycle instead.

If a free piston engine were used to pump gases, say to drive a jackhammer, it could be made simpler, with the air pumped into the combustion chamber and the gas driven out the engine exhaust. The advantage of a free piston engine is obvious: it is much simpler and uses fewer parts than a comparable Diesel engine driving a separate pump.

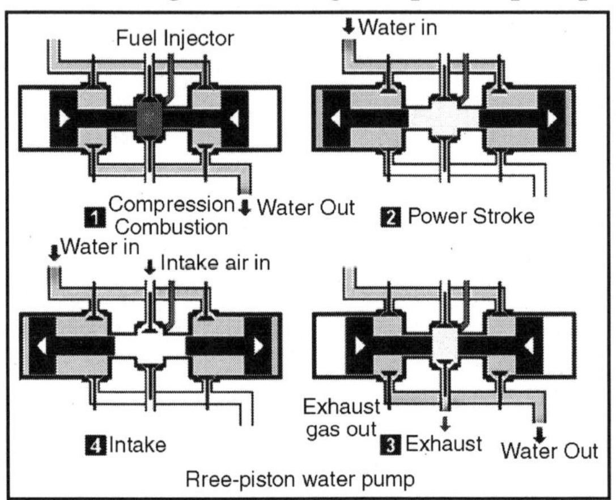

Rree-piston water pump

- Another class of heat engine in common use is the "turbine" engine. The most common application of the turbine engine is in aircraft propulsion, in the form of "turbojet", "turbofan", and "turboshaft" engines, but it is also used to propel ships and to provide electrical power generation.

Turbine engines are generally known as "Brayton cycle" engines. A turbojet is an "open" Brayton cycle engine, in that it operates on a continuous flow basis rather than in a loop, as follows:

- Intake air is first compressed by a "compressor" assembly. This is an adiabatic process.
- Fuel is then mixed with the compressed air and burned in the "combustion chambers" of the engine. This is a constant-pressure process.
- The hot gases generated in the combustion chamber flow out the back to provide reaction thrust, and also spin a "turbine" to run the compressor.

A turbojet is a little different from the other engines considered in this section because it is mainly intended to generate thrust. Turbine engines used to power ground vehicles, "turboprop" engines used to spin an aircraft propeller, and "turboshaft" engines used to spin a helicopter rotor mainly generate torque.

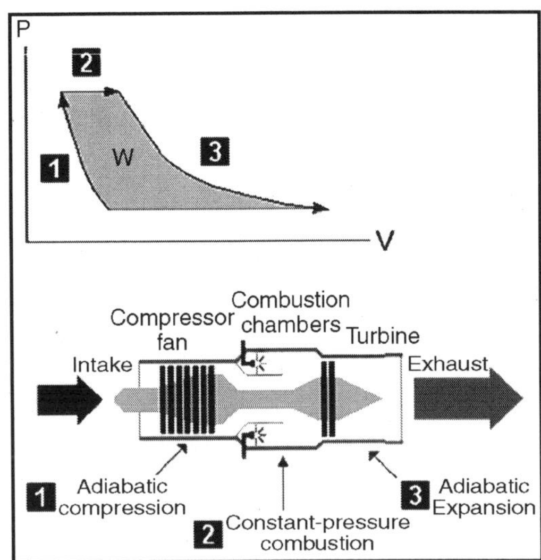

"Closed" cycle turbine engines are used in powerplants and on large vessels. They may operate much like aircraft turboshafts, or they may be driven by compressed steam heated by an external thermal source. The steam then drives a turbine to provide power, and is finally cooled to be routed around to the intake of the compressor again. Closed cycle engines include secondary "heat exchangers" to draw some of the heat out of the exhaust of the turbine and preheat the input, improving energy efficiency.

POWER TURBINE SYSTEM (GVG / PD)

- One elegantly simple example of an open-cycle turbine-based heat engine is a solar power plant that may be built in the hot deserts of Australia. The design looks something like a big chimney, as tall as a skyscraper, that contains a set of turbines. The chimney is open at the bottom and is surrounded by a circular greenhouse-type structure in which the ceiling rises from the rim to the centre. The sun heats up the air in the greenhouse; the hot air flows inward and then into the openings in bottom of the chimney, and rises up the chimney into the cooler air at the top, driving the turbines.

What is particularly interesting about this scheme from the point of view of elementary physics is that it illustrates how a temperature difference, in this case between the bottom and top of the chimney, is required to allow the engine to work.

If it wasn't, the air at the bottom of the chimney would not rise and the turbines would stay idle. In addition, the greater the temperature difference the hotter the air at the bottom of the chimney compared to the air at the top the faster the airflow up the chimney, allowing more power to be obtained out of the turbines.

- Another interesting and unusual engine from a physics point of view is the "dipping bird" toy. This item consists of a bird with a glass body made of a tube with a head and beak on one end and a relatively large bulb on the other. The body pivots on a two-legged stand. The bulb is filled with a fluid that vaporizes easily at room temperature.

To get the dipping bird to work, a glass of water is set in front of it and its beak is placed in the water. The bird will pop upright and stay there for a while; then tip over, put its beak in the water, and remain there for a while; and will then pop upright again, repeating the cycle until the water supply goes dry. What happens is that, while the bird is upright, the volatile fluid in the bulb in the tail evaporates and rises to the head.

This upsets the bird's centre of gravity and it pivots over, dropping its beak into the water. The water cools the fluid and it flows back down to the bulb incidentally, the bird's body never goes horizontal or below horizontal. This shifts the centre of gravity back and the bird goes upright again. The dipping bird obtains its energy from ambient heat, but it needs a cold sink, the glass of water, in order to work. Of course, the bird won't dip if the ambient temperature is too cool to allow the fluid to vaporize, and it won't right itself if the ambient temperature is too warm to allow the fluid to liquify.

This is not a practical engine, though in the 1970s a humour magazine, reflecting the early energy conservation fad of the decade, envisioned giant dipping birds operating as central power stations, and even drew up blueprints. However, the same principle has been considered for practical applications.

In the early 1990s, work was done on the development of torpedo-sized roving robot submarines for oceanographic studies. These machines known as "Slocums" for Joshua Slocum, who in 1898 was the first person to sail around the world solo are powered solely by temperature differences in the oceans.

A Canadian engineer named Doug Webb came up with the idea for the powerplant. He had been working on a sea float that could control its floatation. The float contained an oil reservoir and an external bladder: the oil was pumped out of the float into the bladder to make the thing rise, and pumped back in to make it sink. This was a simple enough idea, but Webb kept thinking it over and coming up with refinements.

First, he decided that he might not really need to use a pump, he could just shift oil around by heating and cooling it. Next, he realized that the temperature differences at different depths in the ocean could do the heating and cooling for him. Finally, he realized that if he put "wings" on the float, it would be able to control its direction as it sank or rose.

The thermal engine that Webb finally came up with hardly looked like an engine at all. The engine components were contained inside a cylindrical pressure casing that made up a body section of the Slocum, except for an inflatable bladder mounted outside the cylinder in the Slocum's frame.

The bladder could be filled with liquid glycol (normally used as automotive antifreeze) and was connected to a matching one inside the cylinder through a tube. The cylinder also contained a pair of pressure tanks, one containing nitrogen and a bellows filled with liquid glycol, and the other containing a special pure hydrocarbon. The components were linked by a system of valves, as shown below:

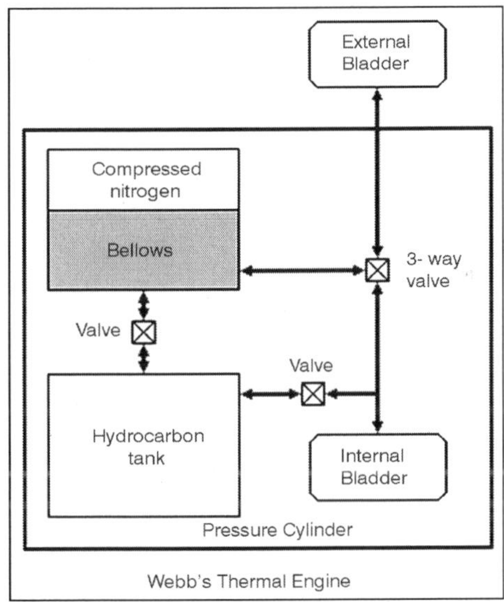

Webb's Thermal Engine

The scheme for operation of this device was devious. The hydrocarbon was liquid at about 15 degrees Celsius and solid at about 5 degrees Celsius, with the transition between the two at about 10 degrees Celsius. Solidified, the hydrocarbon resembled candle wax. On the surface, the external bladder would be filled with glycol; the internal bladder was empty; the hydrocarbon was liquid and the hydrocarbon tank was full; and the bellows was full of glycol. In this state, the Slocum floated. To make it sink, the 3-way valve was opened, causing the glycol to flow into the internal bladder. At the bottom of the Slocum's descent, a kilometre and a half under the sea, the submarine was cold enough to cause the hydrocarbon to freeze. This caused it to shrink by 10 per cent, with the evacuated space drawing off the glycol from the internal bladder through a valve.

Wind Turbine (GVG/PD)

To ascend, the 3-way valve was closed to connect the bellows and the external bladder, and the compressed nitrogen then drove the glycol into the external bladder, causing the Slocum to rise. This done, the 3-way valve was closed again, and the valve between the bellows and the hydrocarbon tank opened.

As the Slocum rose, the hydrocarbon expanded, forcing the glycol back into the bellows and compressing the nitrogen. By the time, the Slocum had reached the surface and was ready for another dive.

The engine could operate effectively under the pressures of the deep sea. No fuel was required, though the Slocum did need battery power to activate the valves. It wasn't a very powerful engine: in practice, it could propel a 40 kilogram Slocum no faster than a kilometre per hour.

That was still fast enough to allow the Slocum to travel across the seas over a period of months or years. Since the Slocum was supposed to operate unattended for five years in a harsh oceanic environment, the diving fins were fixed at an angle for simplicity, with the Slocum adjusting its orientation by shifting the battery pack inside the pressure cylinder.

- Webb got his thermal engine to work, and development of Slocums has continued. A prototype named "Spray" was sent on a journey from Greenland to Spain in the spring of 2006, and plans were made for mass production of the vehicles. It is certainly fun to see such a clever way of getting something for seemingly nothing.

Many forms of renewable energy systems wind power and hydropower have a similar sort of appeal. Obviously, none of these machines actually violate the law of conservation of energy, they just make use of energy sources available in the ambient environment, with the energy inputs provided ultimately by the light and heat of the Sun. Trying to trace out the engine cycles for wind power (global weather) and hydropower (global rain cycle) would be an interesting exercise.

SPECIFIC HEAT CAPACITIES OF IDEAL GAS

If we are dealing with a gas, it is most convenient to use forms of the thermodynamics equations based on the enthalpy of the gas. From the definition of enthalpy:

$$h = e + p * v$$

where h in the specific enthalpy, p is the pressure, v is the specific volume, and e is the specific internal energy. During a process, the values of these variables change. Let's denote the change by the Greek letter delta which looks like a triangle. So "delta h" means the change of "h" from state 1 to state 2 during a process.

$$\text{delta } h = \text{delta } e + p * \text{delta } v$$

The enthalpy, internal energy, and volume are all changed, but the pressure remains the same. From our derivation of enthalpy equation, the change of specific enthalpy is equal to the heat transfer for a constant pressure process:

$$\text{delta } h = cp * \text{delta } T$$

where delta T is the change of temperature of the gas during the process, and c is the specific heat capacity. We have added a subscript "p" to the specific heat capacity to remind us that this value only applies to a constant pressure process. The equation of state of a gas relates the temperature, pressure, and volume through a gas constant R . The gas constant used by aerodynamicists is derived from the universal gas constant, but has a unique value for every gas.

$$p * v = R * T$$

If we have a constant pressure process, then:

$$p * \text{delta } v = R * \text{delta } T$$

Now let us imagine that we have a constant volume process with our gas that produces exactly the same temperature change as the constant pressure process that we have been discussing. Then the first law of thermodynamics tells us:

$$\text{delta } e = \text{delta } q - \text{delta } w$$

where q is the specific heat transfer and w is the work done by the gas. For a constant volume process, the work is equal to zero. And we can express the heat transfer as a constant times the change in temperature. This gives:

delta e = cv * delta T

where delta T is the change of temperature of the gas during the process and c is the specific heat capacity.

If we substitute the expressions for "delta e", "p * delta v", and "delta h" into the enthalpy equation we obtain:

cp * delta T = cv * delta T + R * delta T

dividing by "delta T" gives the relation:

cp = cv + R

We can define an additional variable called the specific heat ratio, which is given the Greek symbol "gamma", which is equal to cp divided by cv:

gamma = cp / cv

"Gamma" is just a number whose value depends on the state of the gas. For air, gamma = 1.4 for standard day conditions. "Gamma" appears in many fluids equations including the equation relating pressure, temperature, and volume during a simple process, the equation for the speed of sound, and all the equations for isentropic flows, and shock waves.

P – V –T SURFACE OF AN IDEAL GAS

For a fixed number of molecules the Ideal Gas Law forms the surface of a three-dimensional graph where the axis are Pressure, Volume, and Temperature.

Lines of constant pressure, constant volume, and constant temperature form a coordinate system labeling the location of an ideal gas.

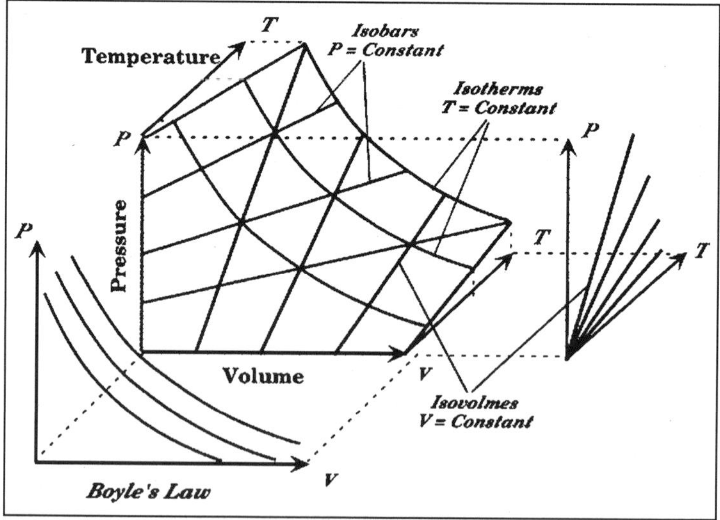

Boyle law showed that the pressure of a low-density gas is inversely proportional to its volume, when the temperature is constant.

$P \alpha 1/V$ for constant temperature.

Charles and Gay-Lussac law showed that the pressure of a low-density gas is proportional to its temperature, when the volume is constant.

$$P \propto T \text{ for constant volume.}$$

The ideal gas law is only valid for low-density gas. There exists a PVT surface for high-density gases only.

PVT SURFACE FOR SUBSTANCE WHICH CONTRACTS ON FREEZING

The equilibrium state of a simple, compressible substance can be specified in terms of its pressure, volume and temperature. If any two of these state variables is specified, we can determine the third. The states of the substance can be represented as a surface in three dimensional PVT space. PVT surface above represents a substance which contracts on freezing. PVT surface image is shown in figure.

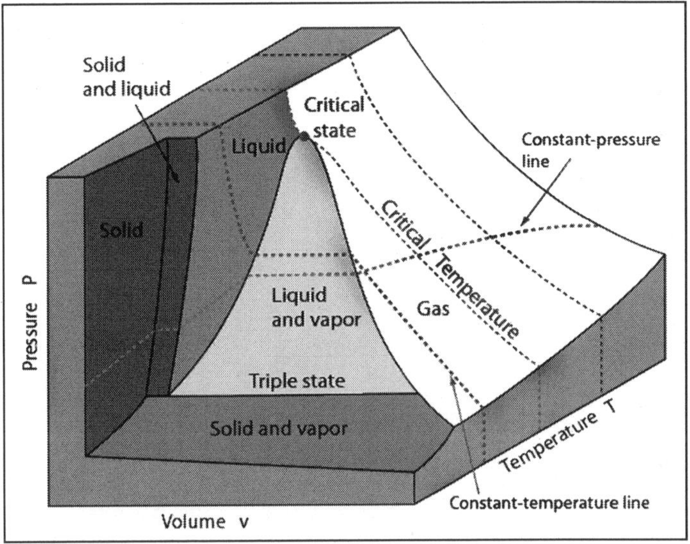

In this the solid, liquid and gas phases can be determined by regions on the surface. A point lying on a line between a single-phase and a two-phase region represents a "saturation state". The line between the liquid and liquid-vapor regions is called as liquid-saturation line. A point on the boundary between the vapor and liquid-vapor regions is called as a saturated-vapor state.

8

Engineering Thermodynamics

DEFINITION OF ENGINEERING THERMODYNAMICS

Engineering Thermodynamics is the study of energies and energy transfer in the systems. It consists of methods and constructions that are used to "account" for macroscopic energy transfer. In fact energy accounting is an appropriate synonym for thermodynamics. In much the same way that accountants balance money in and money out of a bank account, rocket scientists simply balance the energy in and out of a rocket engine. Of course just as a bank account's balance is obfuscated by arcane devices such as interest rates and currency exchange, so too is thermodynamics clouded with seemingly difficult concepts such as irreversibility and enthalpy. But, also just like accounting, a careful review of the rules suggests a coherent strategy for maintaining tabs on a particular account. If a statement about the simplicity of thermodynamics failed to convert would-be students, they may be captured with a few words on the importance of understanding energy transfer in our society. Up until about 150 years ago or so, the earth's economy was primarily fuelled by carbohydrates. That is to say, humans got stuff done by converting food, through a biological process; to fuel we could spend to do work. This was a hindrance to getting things accomplished because, as it turned out, most individuals had to use the brunt of that energy to grow and cultivate more carbohydrates.

Today, we have the luxury, primarily through an understanding of energy, to concentrate our energy production into efficient low maintenance operations. Massive power plants transfer energy to power tools for raising barns. Extremely efficient rocket engines tame and direct massive amounts of energy to blast TV satellites into orbit. This improvement in energy mastery frees humanity's time to engage in more worthwhile activities such as watching cable TV. Understanding energy transfer and energy systems is the next step to destroying the limits to what humanity can next accomplish. The first step is commanding an interest in doing so from an inclined portion of the population.

APPLICATIONS OF ENGINEERING THERMODYNAMICS

SINGLE COMPONENT SYSTEM

All materials can exist in three phases as *solid, liquid,* and *gas.* All one component systems share certain characteristics, so that a study of a typical one component system will be quite useful.

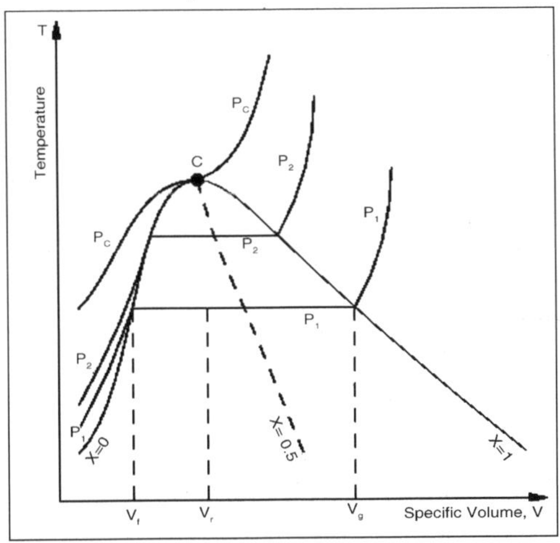

For this consider the above figure, where the heat is transferred to the substance at constant pressure. In this there are temperature and volume curves with all three pressures constants. In this the three line-curves labelled as p_1, p_2, and p are known as isobars which shows conditions at constant pressure. When the liquid and vapour coexist, it is called a saturated state. There is no change in temperature or pressure when liquid and vapour are in equilibrium, so that the temperature is called *saturation temperature* and the pressure is called *saturation pressure.* Saturated states are represented by the horizontal lines in the chart. In the temperature range where both liquid and vapour of a pure substance can coexist in equilibrium, for every value of saturated temperature, there is only one corresponding value of saturation pressure. If the temperature of the liquid is lower than the saturation temperature, it is called *subcooled liquid.* If the temperature of the vapour or gas is greater than the saturation temperature it is called *superheated vapour.*

The amount of liquid and vapour in a saturated mixture is specified by its *quality* x, which is the fraction of vapour in the mixture. Thus, the horizontal line representing the vaporization of the fluid has a quality of $x = 0$ at the left endpoint where it is pure liquid and a quality of $x = 1$ at the right endpoint where it is pure vapour. The blue curve in the preceding diagram shows saturation temperatures for saturated liquid *i.e.* where $x = 0$. The green curve in the diagram shows saturation temperatures for saturated vapour *i.e.* where $x = 1$ which are not isobars.

$$x = \frac{m_g}{m_f + m_g}$$

$$v_x = (1-x)v_f + xv_g$$

$$v_{fg} = v_g - v_f$$

We can say that the point where the solid, liquid and vapour state exist in equilibrium is called as the *triple point*. We further see that as the saturation temperatures increase, the liquid and vapour specific volumes approach each other until the blue and green curves come together and meet at point C on the p_c isobar. At that point C, called the *critical point*, the liquid and vapour states merge together and all their thermodynamic properties become the same. The critical point has a certain temperature T_c, and pressure p_c, which depend on the substance in question. At temperatures above the critical point, the substance is considered a super-heated gas.

HETEROGENEOUS SYSTEM IN EQUILIBRIUM

Gibbs Phase Rule

Gibbs phase rule states that for a heterogeneous system in equilibrium with *C* components in *P* phases, the degree of freedom is given by:

$$F = C - P + 2.$$

Thus, for a single component system with two phases, there is only one degree of freedom *i.e.*: F = 1 – 2 + 2 F = 1 That is, if you are given either the pressure or temperature of wet steam, you can obtain all the properties, while for superheated steam, which has just one phase, you will need both the pressure and the temperature.

Psychrometry

Psychrometry is the study of air and water vapour mixtures which are used for air conditioning. For this the air is taken as a mixture of nitrogen and oxygen with other gases being small enough so that they can be approximated by more of nitrogen and oxygen without much error. In psychrometry, the vapour refers to water vapour. For air at normal (atmospheric) pressure, the saturation pressure of vapour is very low. Also, air is far away from its critical point in those conditions. Thus, the air vapour mixture behaves as an ideal gas mixture. If the partial pressure of the vapour is smaller than the saturation pressure for water for that temperature, the mixture is called unsaturated. The amount of moisture in the air vapour mixture is quantified by its *humidity*.

The absolute humidity ω is the ratio of masses of the vapour and air, *i.e.*, $\omega = m_v/m_a$. Now, applying ideal gas equation, $pV = mRT$ for water vapour and for air, we have, since the volume and temperature are the same, $\omega = 0.622\, p_v/$

p_a. The ratio of specific gas constants (R in preceding equation) of water vapour to air equals 0.622. The *relative humidity* φ is the ratio of the vapour pressure to the saturation vapour pressure at that temperature,

i.e., $\varphi = p_v/p_{v,sat}$.

The *saturation ratio* is the ratio of the absolute humidity to the absolute humidity at saturation, or, $\psi = \omega/\omega_{sat}$. It is easy to see that the saturation ratio is very close to the value of relative humidity.

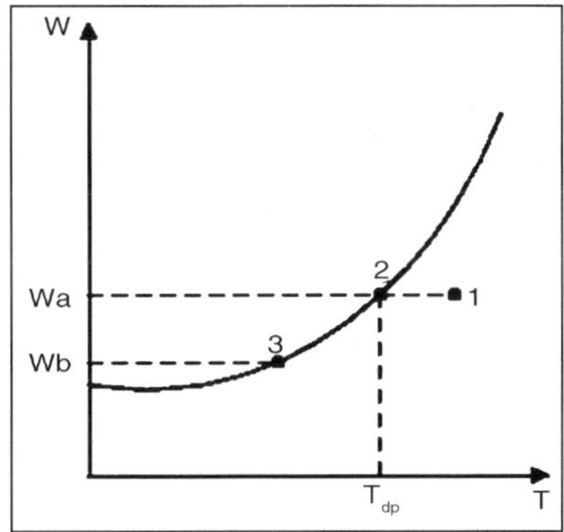

The above figure shows the value of absolute humidity versus the temperature. The initial state of the mixture is 1, and it is cooled isobarically, and at constant absolute humidity. When it reaches 2, it is saturated, and its absolute humidity is ω_a. Further cooling cause's condensation and the system moves to point 3, where its absolute humidity is ω_b. The temperature at 2 is called the *dew point*.

It is customary to state all quantities in psychrometry per unit mass of dry air. Thus, the amount of air condensed in the above chart when moving from 2 to 3 is $\omega_b - \omega_a$.

Adiabatic Saturation

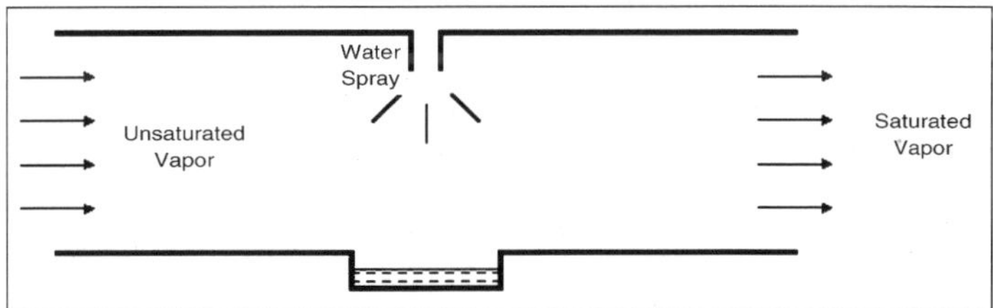

Consider an unsaturated mixture entering a chamber as shown in figure. Suppose water was sprayed into the stream, so that the humidity increases and it leaves as a saturated mixture. This is accompanied by a loss of temperature due to heat being removed from the air which is used for vaporization. If the water supplied is at the temperature of exit of the stream, then there is no heat transfer from the water to the mixture. The final temperature of the mixture is called *adiabatic saturation temperature.*

Psychrometric Chart

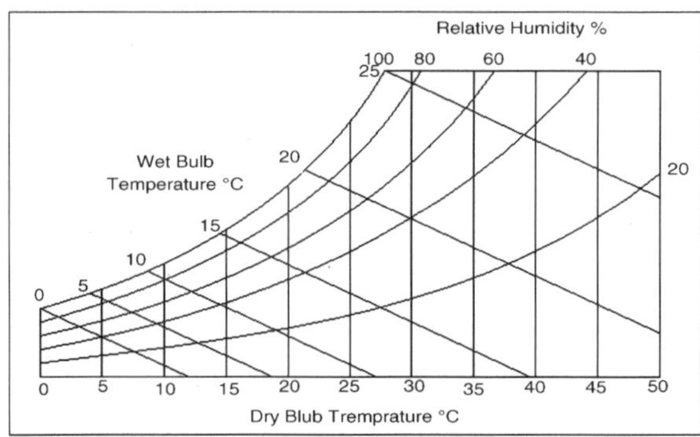

This chart gives the value of absolute humidity versus temperature, along with the enthalpy. From this chart you can determine the relative humidity given the dry and wet bulb temperatures. We have, from the first law, that for a flow system with no heat transfer, the enthalpy is a constant. Now, for the adiabatic saturation process, there is no heat transfer taking place, so that the adiabatic saturation lines are the same as the wet bulb temperature and the constant enthalpy lines.

Air Conditioning

The human body can work efficiently only in a narrow range of conditions. Further, it rejects about 60 W of heat continuously into the surroundings, and more during heavy exercise. The temperature of the body is maintained by the evaporation of sweat from the body. Thus, for comfort, both the temperature and the relative humidity should be low.

Conventional air conditioning consists of setting the humidity at an acceptable level, while reducing the temperature. Reducing the humidity to zero is not the ideal objective. For instance, low humidity leads to issues like high chances of static electricity building up, leading to damage of sensitive electronic equipment. A humidity level of 50% is more acceptable in this case.

The most common method of reducing humidity is to cool the air using a conventional air conditioner working on a reversed Carnot cycle. The vapour that condenses is removed. Now, the air that is produced is very cold, and needs to be heated back up to room temperature before it is released back to the air conditioned area.

Wet Bulb Temperature

The relative humidity of air vapour mixtures is measured by using dry and wet bulb thermometers. The dry bulb thermometer is an ordinary thermometer, while the wet bulb thermometer has its bulb covered by a moist wick. When the mixture flows past the two thermometers, the dry bulb thermometer shows the temperature of the stream, while water evaporates from the wick and its temperature falls. This temperature is very close to the adiabatic saturation temperature if we neglect the heat transfer due to convection.

AN IDEALIZATION OF THERMODYNAMIC CYCLES

THE OTTO CYCLE

The Otto cycle is an idealization of a set of processes used by spark ignition internal combustion engines (2-stroke or 4-stroke cycles).

These engines a) ingest a mixture of fuel and air, b) compress it, c) cause it to react, thus effectively adding heat through converting chemical energy into thermal energy, d) expand the combustion products, and then e) eject the combustion products and replace them with a new charge of fuel and air.

Representation of heat engine as a thermodynamic cycle.

(Ingest mixture of fuel and air)

1′ - 2 Compress mixture quasi-statically and adiabatically

2 - 3 Ignite and burn mixture at constant volume (heat is added)

3 - 4 Expand mixture quasi-statically and adiabatically

4 - 1″ Cool mixture at constant volume (then repeat)

work ~ Force* distance ~ $\int p dv$

fuel use ~ heat added ~ $T_3 - T_2$

efficiency ~ work out/fuel use

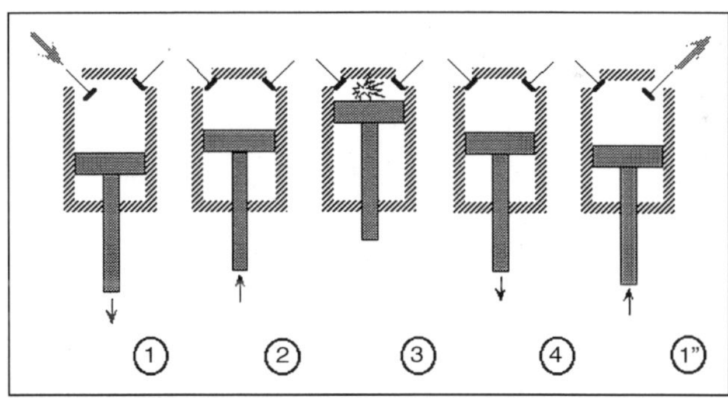

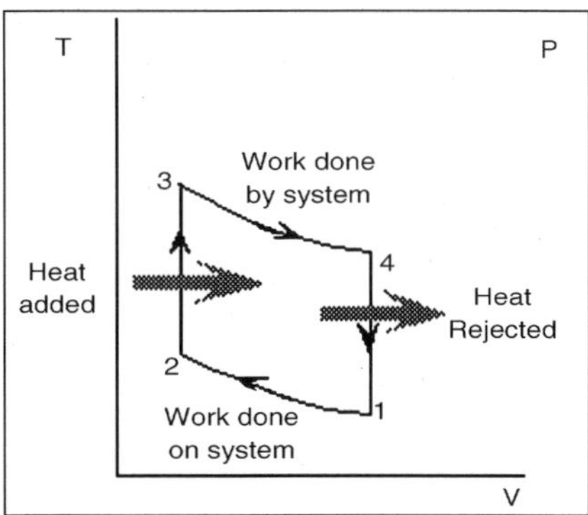

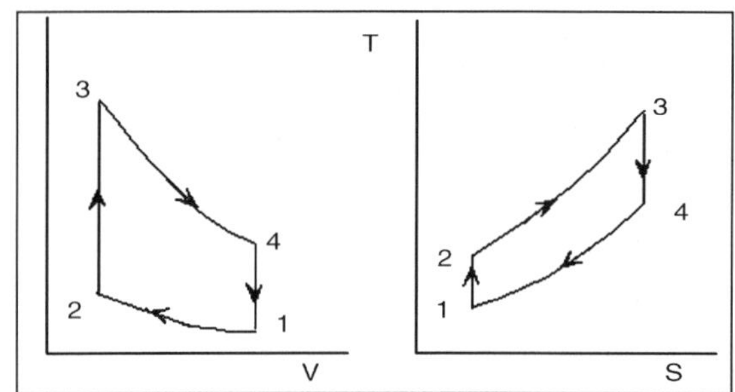

Method for estimating thermal efficiency & work output (First Law of Thermodynamics) Net work done by system = work of expansion + work of compression (–) Both expansion and compression are adiabatic so

$$Dw = (u_3 - u_4) - (u_2 - u_1)$$

Assuming an ideal gas with constant c_v

$$Dw = c_v[(T_3 - T_4) - (T_2 - T_1)]$$

While the above expression is accurate, it is not all that useful. We would like to put the expression in terms of the typical design parameters: the compression ratio ($v_1/v_2 = v_4/v_3$), and the heat added during combustion ($q_{comb.} = c_v(T_3 - T_2)$).

For a quasi-static, adiabatic process,

$$\frac{T_2}{T_1} = \left(\frac{v_1}{v_2}\right)^{\gamma-1} = \frac{T_3}{T_4}$$

so we can write the net work as

$$\Delta w = c_v T_1\left(\frac{T_4}{T_1} - 1\right)\left(r^{\gamma-1} - 1\right)$$

we also know that

$$\frac{T_4}{T_1} = \left(\frac{T_4}{T_3}\right)\left(\frac{T_3}{T_2}\right)\left(\frac{T_2}{T_1}\right) = \frac{T_3}{T_2} \text{ since } \frac{T_2}{T_1} = \frac{T_3}{T_4}$$

so

$$\frac{T_4}{T_1} - 1 = \frac{T_3}{T_2} - 1 = \left(\frac{T_3 - T_2}{T_1}\right)\left(\frac{1}{r^{\gamma-1}}\right)$$

and finally, the result in terms of typical design parameters are:

$$\Delta w = q_{comb.}\frac{\left(r^{\gamma-1} - 1\right)}{r^{\gamma-1}}$$

The thermal efficiency of the cycle is

$$\eta = \frac{\text{network}}{\text{heat input}} = \frac{\Delta w}{q_{comb.}}$$

$$\eta_{\text{Otto}} = \frac{\left(r^{\gamma-1}-1\right)}{r^{\gamma-1}}$$

so

(Rewrite this expression as

$$\eta_{\text{Otto}} = 1-\frac{T_1}{T_2}$$

showing that the efficiency of an Otto cycle depends only on the temperature ratio of the compression process.).

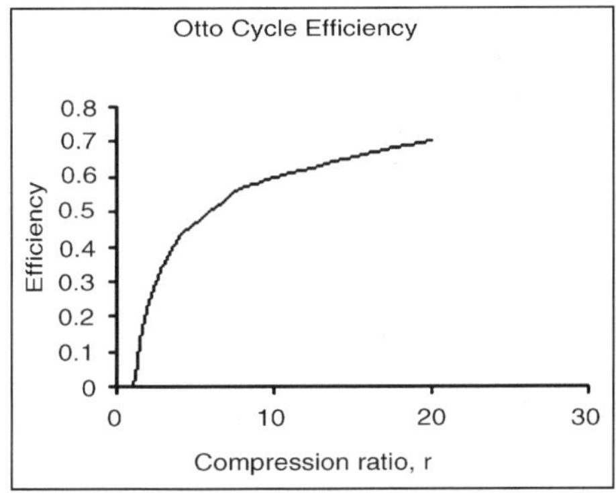

THE SURROUNDINGS AND TRANSFER HEAT

Refrigeration Cycles

Refrigeration cycles take in work from the surroundings and transfer heat from a low temperature reservoir to a high temperature reservoir. Schematically, they look like the diagram given above, but with the direction of the arrows reversed. They can also be recognized on thermodynamic diagrams as closed loops with a counter-clockwise direction of travel. A detailed physical description is given in figure.

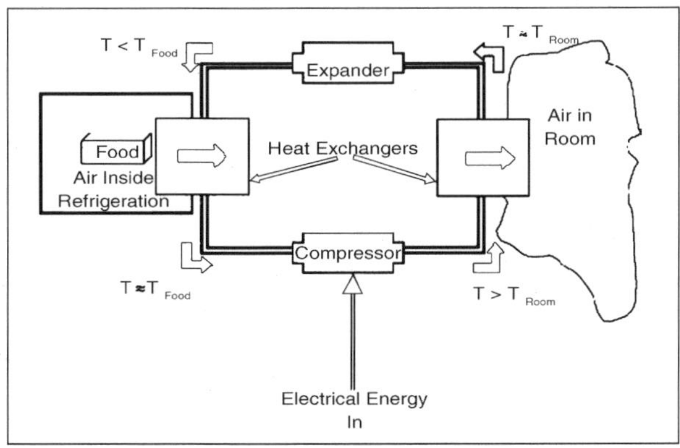

The objective of a refrigerator is to lower the internal energy of a body at low temperature and transfer that energy to the higher temperature surroundings. It requires work to do this. The medium for the energy exchange is a working fluid that circulates in a loop through a series of devices. These devices act to add and remove energy from the working fluid. Typically the working fluid in the loop is considered the thermodynamic system. Sometimes the fluid used alternates between gas-phase and liquid-phase, but this detail is not important for understanding the basic process. A simplified diagram is shown in figure.

As the fluid circulates around the loop, its internal energy is alternately raised and lowered by a series of devices. In this manner, the working fluid is colder than the refrigerator air at one point and hotter than the air in the room at another point. Thus heat will flow in the appropriate direction as shown by the two arrows in the heat exchangers. Starting in the upper right hand corner of the diagram, first the internal energy is lowered either by passing through a small turbine or through an expansion valve. In these devices, work is done by the refrigerant so its internal energy is lowered.

The internal energy is lowered to a point where the temperature of the refrigerant is lower than that of the air in the refrigerator. A heat exchanger is used to transfer energy from the air (and food) in the refrigerator to the cold refrigerant (energy transferred by virtue of a temperature difference only = heat). This lowers the internal energy of the air/food and raises the internal energy of the refrigerant. Then a pump or compressor is used to do work on the refrigerant adding additional energy to it and thus further raising its internal energy. Electrical energy is used to drive the pump or compressor.

The internal energy of the refrigerant is raised to a point where its temperature is hotter than the temperature of the room. The refrigerant is then passed through a heat exchanger (the coils at the back of the refrigerator) so that energy is transferred from the refrigerant to the surroundings. As a result, the internal energy of the refrigerant is reduced and the internal energy of the surroundings is increased. It is at this point where the internal energy of the food and the energy used to drive the compressor or pump are transferred to the surroundings. The refrigerant then continues on to the turbine, repeating the cycle.

THERMODYNAMIC PROCESSES

Brayton Cycle

Brayton cycle is an ideally a set of thermodynamic processes, which is used in turbine engines, for jet propulsion or for generation of electrical power. Components of a gas turbine engine are shown in figure.

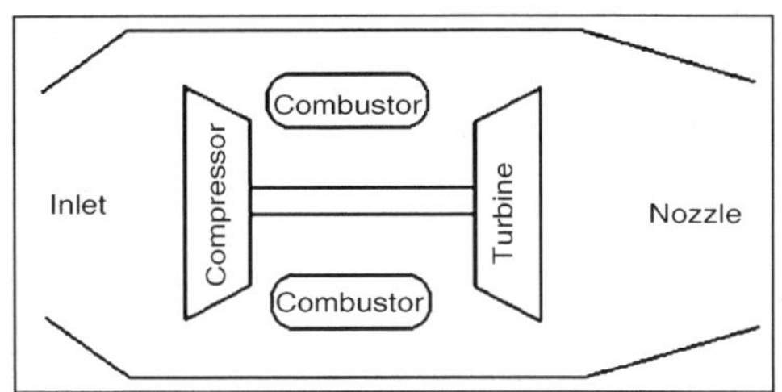

Schematics of typical gas turbine engine: J57 turbojet with afterburning is shown in figure.

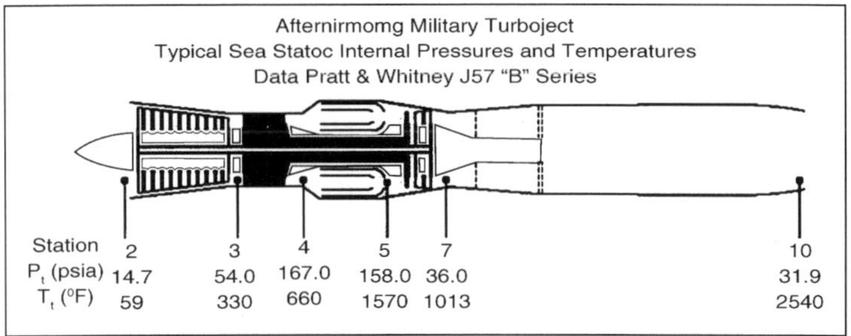

Process of Thermodynamic cycle

The cycle consists of four processes:

- Quasi-static adiabatic compression in the inlet and compressor
- Constant pressure heat addition in the combustor
- Quasi-static adiabatic expansion in the turbine and exhaust nozzle, and finally
- Constant pressure cooling to get the working fluid back to the initial condition.

1-2 Adiabatic, quasi-static compression in inlet and compressor

2-3 Combust fuel at constant pressure (*i.e.* add heat)

3-4 Adiabatic, quasi-static expansion in turbine

- Take work out and use it to drive the compressor
- Take remaining work out and use it to accelerate fluid for jet propulsion, or to turn a generator for electrical power generation.

4-1 Cool the air at constant pressure

Then repeat

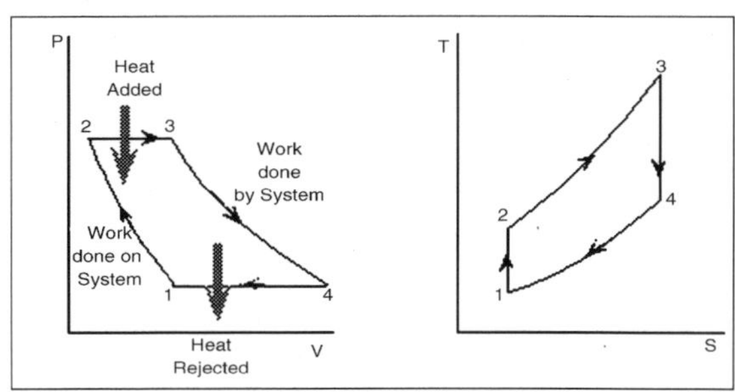

Estimating the Performance of the Engine

First to derive expressions for the net work and the thermal efficiency of the cycle, and then to manipulate these expressions to put them in terms of typical design parameters so that they will be more useful.

First, remember from the First Law we can show that for any cyclic process heat and work transfers are numerically equal

$$Du = q - w$$

$$u_{\text{final}} = u_{\text{initial}} \text{ therefore } Du = 0 \text{ and}$$

$$q = w \text{ or } \oint \delta q = \oint \delta w$$

This fact is often useful for solving thermodynamic cycles. For instance in this example, we would like to find the net work of the cycle and we could calculate this by taking the difference of the work done all the way around the cycle. Or, since Dq = Dw, we could just as well consider the difference between the heat added to the cycle in process 2-3, and the heat rejected by the cycle in process 4-1.

First Law in terms of enthalpy for an ideal gas undergoing a quasi-static process:

$$dq = dh - vdp \text{ or } dq = c_p dT - vdp$$

at constant pressure

$$q_{\text{added}} = c_p\, DT \text{ or } q_{\text{added}} = c_p\, (T_3 - T_2)$$

Hat rejected between 4-1:

Similarly

$$q_{\text{added}} = c_p\, DT \text{ or } q_{\text{rejected}} = c_p\, (T_4 - T_1)$$

Work done and thermal efficiency:

$$Dw = Dq = q_{\text{added}} - q_{\text{rejected}}$$

$$= c_p[(T_3 - T_2) - (T_4 - T_1)]$$

$$h_{\text{Brayton}} = (q_{\text{added}} - q_{\text{rejected}})/q_{\text{added}}$$

$$= [(T_3 - T_2) - (T_4 - T_1)]/(T_3 - T_2)$$

Again, while these expressions are accurate, they are not all that useful. We need to manipulate them to put them in terms of typical design parameters for gas turbine engines. For gas turbine engines the most useful design parameters to use for these equations are often the inlet temperature (T_1), the compressor pressure ratio (p_2/p_1), and the maximum cycle temperature, the turbine inlet temperature (T_3).

Rewriting equations in terms of design parameters:

$$\Delta w = c_p T_1 \left(\frac{T_3}{T_1} - \frac{T_2}{T_1} - \frac{T_4}{T_1} + 1 \right)$$

$$\frac{P_2}{P_1} = \frac{P_3}{P_4} = \left(\frac{T_2}{T_1} \right)^{\gamma/\gamma-1} = \left(\frac{T_3}{T_4} \right)^{\gamma/\gamma-1} \text{ therefore}$$

$$\frac{T_2}{T_1} = \frac{T_3}{T_4} \text{ and } \frac{T_4}{T_1} = \frac{T_3}{T_2}$$

so

$$\Delta w = C_P T_1 \left[\frac{T_3}{T_1} - \left(\frac{P_2}{P_1} \right)^{\gamma-1/\gamma} - \frac{T_3}{T_2} + 1 \right]$$

with

$$T_2 = T_1 \left(\frac{P_2}{P_1} \right)^{\gamma-1/\gamma}$$

becomes

$$\Delta w = C_P T_1 \left[\frac{T_3}{T_1} - \left(\frac{P_2}{P_1} \right)^{\gamma-1/\gamma} - \frac{T_3}{T_1} \left(\frac{P_2}{P_1} \right)^{-(\gamma-1)/\gamma} + 1 \right]$$

and for the efficiency

$$\eta = 1 - \left(\frac{T_4 - T_1}{T_3 - T_2} \right) = 1 - \left[\frac{\frac{T_4}{T_1} - 1}{\frac{T_3}{T_2} - 1} \right]$$

but from above

$$\frac{T_4}{T_1} = \frac{T_3}{T_2}$$

so

$$h_{Brayton} = 1 - \frac{T_1}{T_2} = 1 - \frac{1}{\left(\frac{P_2}{P_1} \right)^{\gamma-1/\gamma}}$$

Performance plots of Brayton Cycle

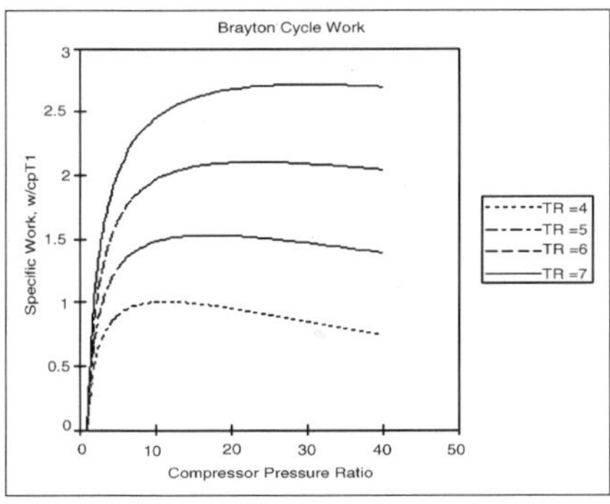

In the plot above, TR = T_3/T_1. For a given turbine inlet temperature, T_3, there is a compressor pressure ratio that maximizes the work.

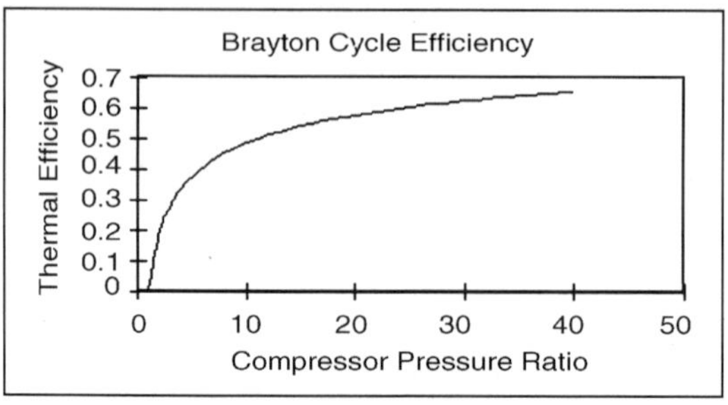

Generalized Representation of Thermodynamic Cycles

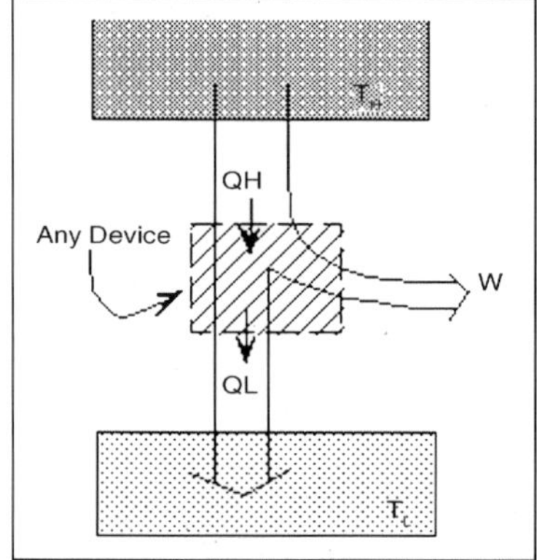

We see that the heat engines can be represented generally as a transfer of heat from a high temperature reservoir to a device + a rejection of heat from a device to a low temperature reservoir + net work done on surroundings

9

Statistical Thermodynamics

Let us briefly review the material which we have covered so far in this course. We started off by studying the mathematics of probability. We then used probabilistic reasoning to analyze the dynamics of many particle systems, a subject area known as statistical mechanics. Next, we explored the physics of heat and work, the study of which is termed thermodynamics. The final step in our investigation is to combine statistical mechanics with thermodynamics: in other words, to investigate heat and work via statistical arguments. This discipline is called statistical thermodynamics, and forms the central subject matter of this course.

MECHANICAL INTERACTION AND MACROSYSTEMS

Let us now examine a purely mechanical interaction between macrostates, where one or more of the external parameters is modified, but there is no exchange of heat energy. Consider, for the sake of simplicity, a situation where only one external parameter x of the system is free to vary. In general, the number of microstates accessible to the system when the overall energy lies between E and $E + \delta E$ depends on the particular value of x, so we can write $\Omega \equiv \Omega(E, x)$. When x is changed by the amount dx, the energy $E_r(x)$ of a given microstate r changes by $(\partial E_r/\partial x)dx$. The number of states $\sigma(E, x)$ whose energy is changed from a value less than E to a value greater than E when the parameter changes from x to $x + dx$ is given by the number of microstates per unit energy range multiplied by the average shift in energy of the microstates. Hence,

$$\sigma(E,x) = \frac{\Omega(E,x)}{\delta E}\overline{\frac{\partial E_r}{\partial x}}dx,$$

where the mean value of $\partial E_r/\partial x$is taken over all accessible microstates (i.e., all states where the energy lies between E and $E + \delta E$ and the external parameter takes the value). The above equation can also be written

$$s(E, x) = -\frac{\Omega(E,x)}{\delta E}\overline{X}\ dx,$$

where

$$\bar{X}(E, x) = -\overline{\frac{\partial E_r}{\partial x}}$$

is the mean generalized force conjugate to the external parameter x.

Consider the total number of microstates between E and $E + \delta E$. When the external parameter changes from to $x + dx$, the number of states in this energy range changes by $(\partial \Omega/\partial x)\, dx$. This change is due to the difference between the number of states which enter the range because their energy is changed from a value less than E to one greater than E and the number which leave because their energy is changed from a value less than $E + \delta E$ to one greater than $E + \delta E$. In symbols,

$$\frac{\partial \Omega\,(E, x)}{\partial x} dx = \sigma(E) - \sigma(E + \delta E) \simeq -\frac{\partial \sigma}{\partial E}\,\delta E,$$

which yields

$$\frac{\partial \Omega}{\partial x} = \frac{\partial(\Omega \bar{X})}{\partial E},$$

where use has been made of Eq. Dividing both sides by Ω gives

$$\frac{\partial \ln \Omega}{\partial x} = \frac{\partial \ln \Omega}{\partial E} \bar{X} + \frac{\partial \bar{X}}{\partial E}.$$

However, according to the usual estimate $\Omega \propto E^f$, the first term on the right-hand side is of order $(f/\bar{E})\bar{X}$, whereas the second term is only of order. Clearly, for a macroscopic system with many degrees of freedom, the second term is utterly negligible, so we have

$$\frac{\partial \ln \Omega}{\partial x} = \frac{\partial \ln \Omega}{\partial E} \bar{X} = \beta \bar{X}.$$

When there are several external parameters $x_1,\ldots, x_n$ so that $\Omega \equiv \Omega(E, x_1, \ldots, x_n)$, the above derivation is valid for each parameter taken in isolation. Thus,

$$\frac{\partial \ln \Omega}{\partial x_\alpha} = \beta \bar{X}_\alpha,$$

where $\bar{X}_\alpha$ is the mean generalized force conjugate to the parameter x_α.

General Interaction between Macrosystems

Consider two systems, A and A', which can interact by exchanging heat energy and doing work on one another. Let the system A' have energy E' and adjustable external parameters $x_1,\ldots,x_n$. Likewise, let the system $A^{(0)} = A + A'$have energy and adjustable external parameters $x'_1,\ldots,x'_n$. The combined system $A^{(0)} = A + A'$is assumed to be isolated. It follows from the first law of thermodynamics that

$$E + E' = E^{(0)} = \text{constant.}$$

Thus, the energy E' of system A' is determined once the energy E of system A is given, and vice versa. In fact, E'could be regarded as a function of E. Furthermore, if the two systems can interact mechanically then, in general, the parameters x' are some function of the parameters x. As a simple example, if the two systems are separated by a movable partition in an enclosure of fixed volume $V^{(0)}$, then

$$V + V' = V^{(0)} = \text{constant,}$$

where V and V' are the volumes of systems A and A', respectively.

The total number of microstates accessible to $A^{(0)}$ is clearly a function of E and the parameters x_α (where runs from 1 to n), so $\Omega^{(0)} \equiv \Omega^{(0)}(E, x_1,\ldots,x_n)$. We have already demonstrated that $\Omega^{(0)}$ exhibits a very pronounced maximum at one particular value of the energy $E = \tilde{E}$ when E is varied but the external parameters are held constant. This behaviour comes about because of the very strong,

$$\Omega \propto E^f,$$

increase in the number of accessible microstates of A(or A') with energy. The number of accessible microstates exhibits a similar strong increase with the volume, which is a typical external parameter, so that

$$\Omega \propto V^f.$$

It follows that the variation of $\Omega^{(0)}$ with a typical parameter x_α, when all the other parameters and the energy are held constant, also exhibits a very sharp maximum at some particular value $x_\alpha = \tilde{x}_\alpha$. The equilibrium situation corresponds to the configuration of maximum probability, in which virtually all systems $A^{(0)}$ in the ensemble have values of E and x_α very close to $\tilde{E}$ and $\tilde{x}_\alpha$. The mean values of these quantities are thus given by $\bar{E} = \tilde{E}$ and $\bar{x}_\alpha = \tilde{x}_\alpha$.

Consider a quasi-static process in which the system A is brought from an equilibrium state described by $\bar{E}$ and $\bar{x}_\alpha$ to an infinitesimally different equilibrium state described by and. Let us calculate the resultant change in the number of microstates accessible to A. Since $\Omega \equiv \Omega(E, x_1,\ldots, x_n)$, the change in $\ln \Omega$ follows from standard mathematics:

$$d\ln\Omega = \frac{\partial \ln\Omega}{\partial E} d\bar{E} + \sum_{\alpha=1}^{n} \frac{\partial \ln\Omega}{\partial x_\alpha} d\bar{x}_\alpha .$$

However, we have previously demonstrated that

$$\beta = \frac{\partial \ln\Omega}{\partial E}, \beta\bar{X}_\alpha = \frac{\partial \ln\Omega}{\partial x_\alpha}$$

$$d\ln\Omega = \beta\left(d\bar{E} + \sum_\alpha \bar{X}_\alpha d\bar{x}_\alpha\right).$$

Note that the temperature β parameter and the mean conjugate forces $\bar{X}_\alpha$ are only well-defined for equilibrium states. This is why we are only considering quasi-static changes in which the two systems are always arbitrarily close to equilibrium.

Let us rewrite Eq. in terms of the thermodynamic temperature, T using the relation $\beta \equiv 1/kT$. We obtain

$$dS = \left(d\bar{E} + \sum_{\alpha} \bar{X}_{\alpha} d\bar{x}_{\alpha} \right) / T,$$

where

$$S = k \ln \Omega.$$

Equation is a differential relation which enables us to calculate the quantity S as a function of the mean energy $\bar{E}$ and the mean external parameters $\bar{x}_{\alpha}$, assuming that we can calculate the temperature T and mean conjugate forces $\bar{X}_{\alpha}$ for each equilibrium state. The function $S(\bar{E}, \bar{x}_{\alpha})$ is termed the entropy of system A. The word entropy is derived from the Greek en+trepien, which means "in change." The reason for this etymology will become apparent presently. It can be seen from Eq. that the entropy is merely a parameterization of the number of accessible microstates. Hence, according to statistical mechanics, is essentially a measure of the relative probability of a state characterized by values of the mean energy and mean external parameters $\bar{E}$ and $\bar{x}_{\alpha}$, respectively.

According to Eq. the net amount of work performed during a quasi-static change is given by

$$đW = \sum_{\alpha} \bar{X}_{\alpha} d\bar{x}_{\alpha}.$$

It follows from Eq. that

$$dS \frac{đW + đW}{T} = \frac{đQ}{T}.$$

Thus, the thermodynamic temperature T is the integrating factor for the first law of thermodynamics,

$$đQ = d\bar{E} + đW,$$

which converts the inexact differential $đQ$ into the exact differential dS. It follows that the entropy difference between any two macrostates and i can f be written

$$S_f - S_i = \int_i^f dS = \int_i^f \frac{đQ}{T},$$

where the integral is evaluated for any process through which the system is brought quasi-statically via a sequence of near-equilibrium configurations from its initial to its final macrostate. The process has to be quasi-static because the temperature T, which appears in the integrand, is only well-defined for an equilibrium state. Since the left-hand side of the above equation only depends on the initial and final states, it follows that the integral on the right-hand side is independent of the particular sequence of quasi-static changes used to get from to f. Thus, $\int_i^f đQ/T$ is independent of the process (provided that it is quasi-static).

All of the concepts which we have encountered up to now in this course, such as temperature, heat, energy, volume, pressure, etc., have been fairly familiar to us from other branches of Physics. However, entropy, which turns out to be of crucial importance in thermodynamics, is something quite new. Let us consider the following questions. What does the entropy of a system actually signify? What use is the concept of entropy?

TEMPERATURE

Suppose that the systems *A* and *A′* are initially thermally isolated from one another, with respective energies E_i and $E_i^{'}$. (Since the energy of an isolated system cannot fluctuate, we do not have to bother with mean energies here.) If the two systems are subsequently placed in thermal contact, so that they are free to exchange heat energy, then, in general, the resulting state is an extremely improbable one [i.e., $P(E_i)$is much less than the peak probability]. The configuration will, therefore, tend to change in time until the two systems attain final mean energies $\overline{E}_f$ and $\overline{E}_f^{'}$ which are such that

$$\beta_f = \beta_f^{'},$$

where $\beta_f \equiv \beta(\overline{E}_f)$ and $\beta_f^{'} \equiv \beta'(\overline{E}_f^{'})$. This corresponds to the state of maximum probability. In the special case where the initial energies, E_i and $E_i^{'}$, lie very close to the final mean energies, $\overline{E}_f$ and $\overline{E}_f^{'}$, respectively, there is no change in the two systems when they are brought into thermal contact, since the initial state already corresponds to a state of maximum probability.

It follows from energy conservation that

$$\overline{E}_f + \overline{E}_f^{'} = E_i + E_i^{'}.$$

The mean energy change in each system is simply the net heat absorbed, so that

$$Q \text{ º } \overline{E}_f - E_i,$$

$$Q\text{¢ º } \overline{E}_f^{'} - E_i^{'}.$$

The conservation of energy then reduces to

$$Q + Q' = 0:$$

i.e., the heat given off by one system is equal to the heat absorbed by the other (in our notation absorbed heat is positive and emitted heat is negative).

It is clear that if the systems *A* and *A′* are suddenly brought into thermal contact then they will only exchange heat and evolve towards a new equilibrium state if the final state is more probable than the initial one. In other words, if

$$P(\overline{E}_f) > (E_i),$$

or

$$\ln P(\overline{E}_f) > \ln P(E_i),$$

since the logarithm is a monotonic function. The above inequality can be written

$$\ln \Omega(\bar{E}_f) + \ln \Omega'(\bar{E}_i') > \ln \Omega(E_i) + \ln \Omega'(E_i'),$$

with the aid of Eq. Taylor expansion to first order yields

$$\frac{\partial \ln \Omega(E_i)}{\partial E}(\bar{E}_f - E_i) + \frac{\partial \ln \Omega'(E_l')}{\partial E'}(\bar{E}_f' - E_i') > 0,$$

which finally gives

$$(\beta_i - \beta_i')\,Q > 0,$$

where $\beta_i \equiv \beta(E_i)$, $\beta_i' \equiv \beta'(E_i')$, and use has been made of Eq.

It is clear, from the above, that the parameter β, defined

$$\beta = \frac{\partial \ln \Omega}{\partial E},$$

has the following properties:

If two systems separately in equilibrium have the same value of β then the systems will remain in equilibrium when brought into thermal contact with one another.

If two systems separately in equilibrium have different values of β then the systems will not remain in equilibrium when brought into thermal contact with one another. Instead, the system with the higher value of β will absorb heat from the other system until the two β values are the same.

Incidentally, a partial derivative is used in Eq. because in a purely thermal interaction the external parameters of the system are held constant whilst the energy changes.

Let us define the dimensionless parameter T, such that

$$\frac{1}{kT} \equiv \beta \equiv \frac{\partial \ln \Omega}{\partial E},$$

where k is a positive constant having the dimensions of energy. The parameter T is termed the thermodynamic temperature, and controls heat flow in much the same manner as a conventional temperature. Thus, if two isolated systems in equilibrium possess the same thermodynamic temperature then they will remain in equilibrium when brought into thermal contact.

However, if the two systems have different thermodynamic temperatures then heat will flow from the system with the higher temperature (i.e., the "hotter" system) to the system with the lower temperature until the temperatures of the two systems are the same. In addition, suppose that we have three systems A, B, and C. We know that if A and B remain in equilibrium when brought into thermal contact then their temperatures are the same, so that $T_A = T_B$. Similarly, if B and C remain in equilibrium when brought into thermal contact, then $T_B = T_C$. But, we can then conclude that $T_A = T_C$, so systems A and C will also remain in equilibrium when brought into thermal contact. Thus, we arrive at the following statement, which is sometimes called the zeroth law of thermodynamics:

If two systems are separately in thermal equilibrium with a third system then they must also be in thermal equilibrium with one another.

The thermodynamic temperature of a macroscopic body, as defined in Eq. depends only on the rate of change of the number of accessible microstates with the total energy. Thus, it is possible to define a thermodynamic temperature for systems with radically different microscopic structures (e.g., matter and radiation). The thermodynamic, or absolute, scale of temperature is measured in degrees kelvin. The parameter k is chosen to make this temperature scale accord as much as possible with more conventional temperature scales. The choice

$$k = 1.381 \times 10^{-23} \text{ joules/kelvin,}$$

ensures that there are 100 degrees kelvin between the freezing and boiling points of water at atmospheric pressure (the two temperatures are 273.15 and 373.15 degrees kelvin, respectively). The above number is known as the Boltzmann constant. In fact, the Boltzmann constant is fixed by international convention so as to make the triple point of water (i.e., the unique temperature at which the three phases of water co-exist in thermal equilibrium) exactly 273.16°K. Note that the zero of the thermodynamic scale, the so called absolute zero of temperature, does not correspond to the freezing point of water, but to some far more physically significant temperature which.

The familiar $\Omega \propto E^f$ scaling for translational degrees of freedom yields

$$kT \sim \frac{\overline{E}}{f},$$

using Eq. so kT is a rough measure of the mean energy associated with each degree of freedom in the system. In fact, for a classical system (i.e., one in which quantum effects are unimportant) it is possible to show that the mean energy $(1/2)kT$ associated with each degree of freedom is exactly. This result, which is known as the equipartition theorem.

The absolute temperature T is usually positive, since $\Omega(E)$ is ordinarily a very rapidly increasing function of energy. In fact, this is the case for all conventional systems where the kinetic energy of the particles is taken into account, because there is no upper bound on the possible energy of the system, and $\Omega(E)$ consequently increases roughly like E^f. It is, however, possible to envisage a situation in which we ignore the translational degrees of freedom of a system, and concentrate only on its spin degrees of freedom. In this case, there is an upper bound to the possible energy of the system (i.e., all spins lined up anti-parallel to an applied magnetic field). Consequently, the total number of states available to the system is finite. In this situation, the density of spin states $\Omega_{spin}(E)$ first increases with increasing energy, as in conventional systems, but then reaches a maximum and decreases again. Thus, it is possible to get absolute spin temperatures which are negative, as well as positive.

In Lavoisier's calorific theory, the basic mechanism which forces heat to flow from hot to cold bodies is the supposed mutual repulsion of the constituent particles of calorific fluid. In statistical mechanics, the explanation is far less contrived. Heat flow occurs because statistical systems tend to evolve towards their most probable states, subject to the imposed physical constraints.

When two bodies at different temperatures are suddenly placed in thermal contact, the initial state corresponds to a spectacularly improbable state of the overall system. For systems containing of order 1 mole of particles, the only reasonably probable final equilibrium states are such that the two bodies differ in temperature by less than 1 part in 10^{12}. The evolution of the system towards these final states (i.e., towards thermal equilibrium) is effectively driven by probability.

STATISTICAL MECHANICS

The kinetic theory of matter attempts to explain the empirical regularities occurring in the macroscopic properties of material objects in terms of the microscopic behaviour of their atomic and molecular constituents as described by Newton's laws of mechanics.

The early studies in the revival of this theory which occurred in the second half of the nineteenth century had only limited explanatory power since they assumed that:

- These constituents all had the same velocity; and
- The influence of intermolecular forces and impacts were negligible. The development of the subject took a major leap forward when Maxwell allowed intermolecular collisions to play a role and demonstrated that their effect was to produce a statistical distribution of molecular velocities in which all velocities would occur with a known probability. This is given by Maxwell's famous distribution law:

$$f(x) = \frac{1}{a\sqrt{\pi}} e^{-x^2/a2}$$

where a is a constant and x is the component of the velocity in the x-direction.

Of course, the statistical element was introduced purely as a matter of convenience in order to overcome the practical problems in handling enormously large numbers of atoms and molecules. The latter themselves were regarded as traversing distinct, continuous space-time trajectories and as obeying Newton's laws.

The original proof of the above law was widely regarded as unsatisfactory and subsequent attempts to derive it more rigourously can be divided into three types:

1. If a Maxwellian distribution is already established then it can be shown that conservation of energy implies that further collisions will leave it unchanged. Hence this distribution is the only one which is stable. This was the approach taken by Maxwell himself.

2. One may define a quantity, *H,* which depends on the velocity distribution and then show that the effect of molecular collisions is always to decrease *H,* unless the distribution is Maxwellian, in which case *H* remains constant. This constitutes the essence of Boltzmann's '*H*-Theorem' approach.
3. One may regard the atomic and molecular velocities as random quantities and obtain the best possible estimate of their distribution, subject to the constraints of fixed energy and number of atoms or molecules, using probability theory. Maxwell's derivation may be improved by calculating all the possible ways of dividing the energy among the atoms or molecules, subject to the above constraints. This was the thinking which subsequently developed into the 'Combinatorial Approach' (and which lies behind the permutation argument above).

Approaches II and III rest on very different foundations: the *H*-Theorem Approach is explicitly based on a consideration of molecular trajectories, whereas the Combinatorial Approach eschewed such detailed consideration and was simply concerned with the distribution of molecules over energy states. Historically, the Combinatorial Approach played an absolutely fundamental role in the development of quantum theory.

Maxwell's work had an enormous impact upon Boltzmann, who was seeking a mechanical explanation of the apparent irreversibility of natural processes as expressed by the Second Law of Thermodynamics. Put rather crudely, this states that it is impossible for heat to flow from a colder to a hotter body. The law was expressed by Clausius in 1865 in terms of his newly introduced entropy function *S*, where,

$$dQ \leq T.\, dS$$

where dQ is the heat change, T is the temperature change and dS is the change in entropy. The equality holds for reversible processes, of course. The task facing Boltzmann was then two-fold: first, to demonstrate the existence of an entropy function satisfying Clausius's definition and secondly to show that this function could only increase in an irreversible process. After reading Maxwell's 1867 paper, Boltzmann realised that the key lay with the above distribution function and in 1871 he was able to demonstrate the existence of an entropy function in purely mechanical terms and lay down a procedure for finding it.

The following year he took the next step and analysed the behaviour of his entropy function in irreversible processes by considering the way in which the velocity distribution changed with time due to intermolecular collisions. This work can be understood as both an attempt to complete the statistical mechanical reduction of the Second Law, and as showing that the effect of intermolecular collisions on a gas in a non-equilibrium state would be to drive it to equilibrium as described by Maxwell's law. Boltzmann was able to show that the Maxwellian distribution represents the equilibrium state and obtained an explicit formula for the rate of change of f, $\frac{\partial f}{\partial t}$, on the basis of an "exact consideration of the collision process" between two molecules of a spatially homogenous, low-density gas with no external forces present. With this formula in hand, he was able to show that f *always* tends towards the Maxwellian form. To do this he introduced an auxiliary quantity E (which later came to be called H), defined by,

$$E = \int_0^\infty f(xt)\left\{\log\left[f(x,t\right]t / \sqrt{x} - 1\right\} dx .$$

By considering the symmetrical character of the collision and the possibility of inverse collisions, and assuming, as Maxwell had before him, that the velocities of the two molecules before they collide are statistically independent, Boltzmann demonstrated that E could only decrease with time; that is,

$$\frac{dE}{dz} \leq 0 .$$

With H substituted for E, this corresponds to Boltzmann's H-Theorem. E cannot decrease to infinity, of course, and so the distribution function must approach a form for which E has a minimum value and its derivative vanishes. At this value the function has, and can only have, the Maxwellian form.

Thus $-E$ increases in the irreversible approach to equilibrium and hence behaves like the entropy function. Furthermore, $-E$ was actually proportional to the entropy in the equilibrium state. This implies, as Boltzmann himself indicated, an entirely new approach to proving the Second Law, one that could deal with the increase in entropy in irreversible processes as well as with its existence as an equilibrium state function. In this way, the H-Theorem effectively extended the definition of entropy to non-equilibrium situations and completed the kinetic reduction of the Second Law of Thermodynamics.

Now, although Maxwell's distribution function introduced a certain statistical 'coarse-graining' into the description, the development of the H-Theorem was still based upon consideration of binary collisions between molecules traversing distinct continuous trajectories. Thus, despite the probabilistic elements, the underlying basis was still, of course, Newtonian and deterministic. The tensions this caused within Boltzmann's edifice are well known.

In particular, his conclusion, that the E/H-function always decreased, was disputed by Maxwell, Tait and Thompson and, independently and famously, by Loschmidt, who noted that in any system, "the entire course of events will be retraced if at some instant the velocities of all its parts are reversed". If entropy is a specifiable function of the positions and velocities of the particles of a system, and if that function increases during some particular motion of the system, then reversing the direction of time in the equations of motion will specify a trajectory through which the entropy must decrease. For every possible motion that leads towards equilibrium, there is another, equally possible, that leads away and is therefore incompatible with the Second Law. Loschmidt concluded that if the kinetic theory were true, then the Second Law could not hold universally and thus Boltzmann's proof of the *H*-Theorem could not be correct.

Boltzmann's 1877 response to the Loschmidt 'paradox' is illuminating. He basically admitted that one could not, in fact, prove that entropy increased 'with absolute necessity' and that, according to probability theory, even the most improbable non-uniform distribution is still not absolutely impossible. However, he then claimed that the existence of such improbable entropy-decreasing situations did not contradict the fact that for the overwhelming majority of initial states the entropy could be counted on to increase and that the improbabilities associated with the former case were, for all practical purposes, impossibilities. Boltzmann now focused on the probabilistic aspect of the *H*-Theorem and argued that it followed from this that the number of states leading to a uniform, equilibrium distribution after a certain time must be much larger than the number leading to a non-uniform distribution, since, he claimed, there are infinitely many more uniform states than non-uniform ones. No justification was actually given for this last claim, although Boltzmann intimated where one might be found:

One could even calculate the possibilities of the various states from the ratios of the number of ways in which their distributions could be achieved, which could perhaps lead to an interesting method of calculating thermal equilibrium.

This 'interesting method' was subsequently elaborated later that same year in one of the most significant papers of classical statistical mechanics, entitled 'Probabilistic Foundations of Heat Theory'.

Prior to this Boltzmann had, in a series of papers from 1868 and 1871, re-derived and extended Maxwell's results and, significantly, in 1868, sketched an alternative derivation that was free from any assumptions regarding inter-molecular collisions. Thus he considered the distribution of a fixed amount of energy over a finite number of molecules in such a way that all combinations were equally probable. By regarding the energy as divided into small but finite packets, Boltzmann could treat this as a problem in combinatorial analysis. In this manner he obtained a complicated expression which reduced to Maxwell's law in the limit of an infinite number of molecules and infinitesimal energy elements. This marks the beginnings of his Combinatorial Approach. In the 1877 work, Boltzmann drew on this earlier work and presented a new and radical alternative to the *H*-Theorem approach. In line with his general philosophical attitude towards physical theories this 'Combinatorial Approach' was elaborated through a succession of models of increasing complexity and closer approximation to the physical situation.

The first such model was a highly simplified and explicitly fictional discrete energy model in which he considered a collection of molecules whose individual energies were restricted to a finite, discrete set, with the total energy held fixed. If ω_k is the number of such molecules with energy $k\varepsilon$, then the set ω_0, α_1., ε_p is sufficient to define a particular macro-state (*Zustandverteilung*) of the gas. Boltzmann then noted that such a macro-state could be achieved in many different ways, each of which he called a 'complexion'. In general, if a complexion was specified by a set of numbers, each fixing the energy $k\,\varepsilon$ of the *i*th molecule, then, he wrote, a second complexion belonging to the same macro-state would be achieved by any permutation of the two molecules *i* and *j* which have different energies. Thus a permutation of particles between different energy states gave rise to a different macro-state of the system as a whole.

INVESTIGATION OF STATISTICAL THERMODYNAMICS

THERMAL INTERACTION BETWEEN MACROSYSTEMS

Let us begin our investigation of statistical thermodynamics by examining a purely thermal interaction between two macroscopic systems, A and A', from a microscopic point of view. Suppose that the energies of these two systems are E and E', respectively. The external parameters are held fixed, so that A and A' cannot do work on one another.

However, we assume that the systems are free to exchange heat energy (i.e., they are in thermal contact). It is convenient to divide the energy scale into small subdivisions of width δE. The number of microstates of A consistent with a macrostate in which the energy lies in the range E to $E + \delta E$ is denoted $\Omega(E)$. Likewise, the number of microstates of consistent with a macrostate in which the energy lies between E' and $E' + \delta E$ is denoted $\Omega'(E')$.

The combined system $A^{(0)} = A + A'$ is assumed to be isolated (i.e., it neither does work on nor exchanges heat with its surroundings). It follows from the first law of thermodynamics that the total energy $E^{(0)}$ is constant. When speaking of thermal contact between two distinct systems, we usually assume that the mutual interaction is sufficiently weak for the energies to be additive. Thus,

$$E + E \simeq E^{(0)} = \text{constant}$$

Of course, in the limit of zero interaction the energies are strictly additive. However, a small residual interaction is always required to enable the two systems to exchange heat energy and, thereby, eventually reach thermal equilibrium. In fact, if the interaction between A and A' is too strong for the energies to be additive then it makes little sense to consider each system in isolation, since the presence of one system clearly strongly perturbs the other, and vice versa. In this case, the smallest system which can realistically be examined in isolation is $A^{(0)}$.

According to Eq. if the energy of A lies in the range E to $E + \delta E$ then the energy of A' must lie between $E^{(0)} - E - \delta E$ and $E^{(0)} - E$. Thus, the number of microstates accessible to each system is given by $\Omega'(E)$ and $(E^{(0)} - E)$, respectively. Since every possible state of A can be combined with every possible state of A' to form a distinct microstate, the total number of distinct states accessible to $A^{(0)}$ when the energy of A lies in the range E to $E + \delta E$ is $\Omega^{(0)} = \Omega(E)\, \Omega'(E^{(0)} - E)$.

Consider an ensemble of pairs of thermally interacting systems, A and A', which are left undisturbed for many relaxation times so that they can attain thermal equilibrium. The principle of equal a priori probabilities is applicable to this situation.

According to this principle, the probability of occurrence of a given macrostate is proportional to the number of accessible microstates, since all microstates are equally likely. Thus, the probability that the system A has an energy lying in the range E to $E + \delta E$can be written

$$P(E) = C\, \Omega(E)\, \Omega'\,(E^{(0)} - E),$$

where C is a constant which is independent of E.

The typical variation of the number of accessible states with energy is of the form

$$\Omega \propto E^{f},$$

where f is the number of degrees of freedom. For a macroscopic system f is an exceedingly large number. It follows that the probability $P(E)$in Eq. is the product of an extremely rapidly increasing function of E and an extremely rapidly decreasing function of E. Hence, we would expect the probability to exhibit a very pronounced maximum at some particular value of the energy.

Let us Taylor expand the logarithm of $P(E)$ in the vicinity of its maximum value, which is assumed to occur at $E = \tilde{E}$. We expand the relatively slowly varying logarithm, rather than the function itself, because the latter varies so rapidly with the energy that the radius of convergence of its Taylor expansion is too small for this expansion to be of any practical use. The expansion of ln Ω (E) yields

$$\ln \Omega(E) = \ln \Omega(\tilde{E}) + \beta(\tilde{E})\eta - \frac{1}{2}\lambda(\tilde{E})\eta^2 + \cdots,$$

where

$$\text{h} = E - \tilde{E},$$

$$\text{b} = \frac{\partial \ln \Omega}{\partial E},$$

$$\lambda = \frac{\partial^2 \ln \Omega}{\partial E^2} = -\frac{\partial \beta}{\partial E}.$$

Now, since $E' = E^{(0)} - E$, we have

$$E' - \tilde{E}' = -(E - \tilde{E}) = -\eta.$$

It follows that

$$\ln \Omega'(E') = \ln \Omega'(\tilde{E}') + \beta'(\tilde{E}')(-\eta) - \frac{1}{2}\lambda'(\tilde{E}')(-\eta)^2 + \cdots,$$

where $\beta'$and $\lambda'$are defined in an analogous manner to the parameters β and λ. Equations can be combined to give

$$\ln[\Omega(E)\Omega'(E')] = \ln[\Omega(\tilde{E})\Omega'(\tilde{E}')] + [\beta(\tilde{E}) - \beta'(\tilde{E}')]\eta - \frac{1}{2}[\lambda(\tilde{E}) + \lambda'(\tilde{E}')]\eta^2 + \cdots.$$

At the maximum of ln $[\Omega(E)\Omega'(E')]$ the linear term in the Taylor expansion must vanish, so

$$\beta(\tilde{E}) = \beta(\tilde{E}'),$$

which enables us to determine $\tilde{E}$. It follows that

$$\ln P(E) = \ln P(\tilde{E}) - \frac{1}{2}\lambda_0 \eta^2,$$

or

$$P(E) = P(\tilde{E}) \exp\left[-\frac{1}{2}\lambda_0 (E - \tilde{E})^2\right],$$

where

$$\lambda_0 = \lambda(\tilde{E}) + \lambda'(\tilde{E}').$$

Now, the parameter λ_0 must be positive, otherwise the probability $P(E)$ does not exhibit a pronounced maximum value: i.e., the combined system $A^{(0)}$ does not possess a well-defined equilibrium state as, physically, we know it must. It is clear that $\lambda(\tilde{E})$ must also be positive, since we could always choose for A' a system with a negligible contribution to λ_0, in which case the constraint $\lambda_0 > 0$ would effectively correspond to $\lambda(\tilde{E}) > 0$. [A similar argument can be used to show that $\lambda'(\tilde{E}')$ must be positive.] The same conclusion also follows from the estimate $\Omega \propto E^f$, which implies that

$$\lambda(\tilde{E}) \sim \frac{f}{\tilde{E}^2} > 0.$$

According to Eq. the probability distribution function $P(E)$ is a Gaussian. This is hardly surprising, since the central limit theorem ensures that the probability distribution for any macroscopic variable, such as $\tilde{E}$, is Gaussian in nature. It follows that the mean value of E corresponds to the situation of maximum probability (i.e., the peak of the Gaussian curve), so that

$$\bar{E} = \tilde{E}.$$

The standard deviation of the distribution is

$$\Delta^* E = \lambda_0^{-1/2} \sim \frac{\bar{E}}{\sqrt{f}},$$

where use has been made of Eq. (assuming that system makes the dominant contribution to λ_0). It follows that the fractional width of the probability distribution function is given by

$$\frac{\Delta^* E}{\bar{E}} \sim \frac{1}{\sqrt{f}}.$$

Hence, if A contains 1 mole of particles then $f \sim N_A \simeq 10^{24}$ and $\Delta^* E/\bar{E} \sim 10^{-12}$. Clearly, the probability distribution for E has an exceedingly sharp maximum. Experimental measurements of this energy will almost always yield the mean value, and the underlying statistical nature of the distribution may not be apparent.

THERMODYNAMIC PROPERTIES OF PLATINUM DIATOMICS

Thermodynamic properties of diatomic transition metal compounds are very important for investigating their thermal behaviour. Recently, these properties have been applied in the fabrication of smart devices using intelligent materials for examples using platinum). Transition metal compounds have a wide range of actual and potential applications in materials science because of their relatively high melting points, moderate densities, and resistance to chemical attack.

Platinum-containing compounds have been used in nanoscience and nanotechnology. For example, alloys such as iron-platinum (FePt), are used as nanodots. Platinum itself is used in a wide range of applications, including catalytic converters for cars, fuel cell electrocatalysts, computer technology, optical communication, missile technology, neurosurgery and medical science.

The unique properties of platinum generate a high level of interest among scientists. Due to the presence of unpaired *d*electrons, which have a greater angular momentum than *s* and *p* electrons, more energy is needed for the excitation of Pt in the molecular phase. This requires a high energy excitation device such as a gas discharge laser, which leads to experimental difficulties. Thus the experimental study of platinum diatomics is very challenging and expensive.

Scientific groups such as the Scientific Group Thermodata Europe (SGTE), the Joint Institute of High Temperatures, Russian Academy of Sciences (IVTAN) and, in the U.S.A., the Joint Army-Navy-Air Force (JANAF) Thermochemical Working Group (6) and the National Aeronautics and Space Administration (NASA), are engaged in the critical assessment and compilation of thermodynamic data for different molecular species. An early contribution to the development of thermodynamic properties was made by

Tolman, for diatomic hydrogen. The credit for further development of the subject goes to Hicks and Mitchell, for their work on hydrogen chloride. Giauque and Overstreet implemented the technique suggested by Hicks and Mitchell for the calculation of these properties for hydrogen, chlorine and hydrogen chloride.

They modified the reported theory by using stretching and interaction terms for diatomic molecules. Gordon and Barnes calculated the thermodynamic properties of chlorine (Cl_2), bromine (Br_2), hydrogen chloride (HCl), carbon monoxide (CO), oxygen (O_2) and nitric oxide (NO) molecules. The study of thermodynamic properties of the phosphorous mononitride (PN) molecule was performed by McCallum and Liefer.

The thermodynamic properties of transition metal alloys were reported by Darby.Calculation of the partition function and thermodynamic properties of the rare gas atoms argon, krypton and xenon was performed by Elyutin *et al.* In the domain of theory of thermodynamic properties, Eu reported Boltzmann entropy, relative entropy and their related values. Chandra and Sharma calculated the partition function for carbon monosulfide (CS) and silicon monoxide (SiO) molecules. Recently Uttam and coworkers estimated the thermodynamic properties of potassium monohalides and alkaline earth metal monohydrides using partition function theory.

The data on thermodynamic properties have been reported for a large number of molecules but a survey of the literature reveals that the values of thermodynamic properties for some diatomic molecules are not yet reported accurately. Therefore we have estimated the thermodynamic properties of platinum monohydride (PtH), platinum monocarbide (PtC), platinum mononitride (PtN) and platinum monoxide (PtO) molecules using spectroscopic data and partition function theory. The choice of the temperature range from 100 K to 3000 K is due to the fact that this range of temperatures covers the applications of platinum from biological sciences to high-temperature chemistry and astrophysics. In the present paper, we report the values of thermodynamic properties Gibbs energy (G), enthalpy (H), entropy (S) and specific heat capacity at constant pressure (C_P) for PtH, PtC, PtN and PtO molecules at different temperatures for which these properties are not given in the literature.

METHOD OF CALCULATION

A diatomic molecule is associated with translational, rotational, vibrational and electronic motions. Corresponding to these four types of motions, there are four types of energy: translational, rotational, vibrational and electronic energy. Translational motion is due to the three dimensional movement of a molecule in space. Rotational motion is due to the rotation of the molecule as a whole about an axis passing through the centre of gravity and perpendicular to the internuclear axis. In diatomic molecules, atoms are also able to vibrate relative to each other along the internuclear axis and this is the origin of vibrational motion.

Motion of electrons in one atom is perturbed by electronic and nuclear motion in the other atom. Due to this, reshuffling of orbitals takes place and that generates molecular orbitals. This phenomenon is responsible for electronic motion. The electronic energy, <“ 1 eV to 10 eV, is very high compared to vibrational energy, <“ 10″^{2} eV, rotational energy, <“ 10″^{3}eV, and translational energy, <“ 10″^{22} eV. However, theory shows that below 3000 K molecules are not excited electronically, and electronic motion only plays a significant role above 3000 K. Therefore electronic motion can be neglected below 3000 K.

RESULTS AND DISCUSSION

The calculated thermodynamic properties namely Gibbs energy, enthalpy, entropy and heat capacity at constant pressure, of PtH, PtC, PtN and PtO molecular gases have been estimated from spectroscopic data and are collected inTables. The spectroscopic constants which were used for the calculation of these properties are displayed in Table V (20). PtH is different from PtC, PtN and PtO since it has a $^{2}\Delta$ ground state. Therefore we have incorporated the ground state multiplicity in our calculation for the improvement of the results. We have estimated the thermodynamic properties from theoretical spectroscopic data and experimental spectroscopic data for the PtH molecule as shown inTable VI.

From comparison, it is clear that the Gibbs energy has a maximum deviation of 0.12%, enthalpy has 0.11%, entropy has 0.17% and heat capacity has 0.63% up to 2000 K. Ideally it is assumed that rotational and vibrational motions are independent of each other, but in practice they interact with each other. In the present calculation we include this effect by taking the vibrational-rotational partition function instead of the independent rotational and vibrational partition functions.

This gives more accurate values of thermodynamic properties than the values obtained from individual rotational and vibrational partition functions. A similar approach has been applied for the calculation of thermodynamic properties of monohalides of potassium, and the obtained results were in close agreement with reported values. Accuracy of the data also depends on the vibrational-rotational coupling. If coupling is weak, the stretching constant (α_e) is sufficient for the calculation of thermodynamic properties. If coupling is strong, the incorporation of the vibration-rotation constant ($\tilde{a}_e$) gives more accurate data.

10

Elementary Concepts in Statistics

Statistics is a mathematical science pertaining to the collection, analysis, interpretation or explanation, and presentation of data. It is applicable to a wide variety of academic disciplines, from the physical and social sciences to the humanities. Statistics are also used for making informed decisions. Statistical methods can be used to summarize or describe a collection of data; this is called descriptive statistics. In addition, patterns in the data may be modeled in a way that accounts for randomness and uncertainty in the observations, and then used to draw inferences about the process or population being studied; this is called inferential statistics. Both descriptive and inferential statistics comprise applied statistics. There is also a discipline called mathematical statistics, which is concerned with the theoretical basis of the subject. The idea that statistics branched off from mathematics is a widely held misconception. Some place an undue emphasis on the relationship, but the two disciplines are very different. The purpose of descriptive statistics is to communicate information, while inferential statistics is used to reach conclusions and deductions that possibly explain the data. Both of these together make up applied statistics. There is also a discipline called mathematical statistics, which is concerned with the theoretical basis of the subject.

FUNDAMENTALS OF STATISTICS

By *statistics* we denote the investigation of regularities in apparently non-deterministic processes. An important basic quantity in this context is the "relative frequency" of an "event". Let us consider a repeatable experiment - say, the throwing of a die - which in each instance leads to one of several possible results e - say, $e \equiv$ number of Points = 4. Now repeat this experiment n times under equal conditions and register the number of cases in which the specific result e occurs; call this number $f_n(e)$. The relative frequency of e is then defined as $r(e) \equiv f_n(e)/n$.Following R. von Mises we denote as the "probability " of an event e the expected value of the relative frequency in the limit of infinitely many experiments:

$$p(e) \equiv \lim_{n \to \infty} \frac{f_n(e)}{n}$$

Example: Game die; 100-1000 trials; $e \equiv \{\text{no.of points} = 6\}$, or,

$$e \equiv \{\text{no.of points} \leq 3\}.$$

Now, this definition does not seem very helpful. It implies that we have already done some experiments to determine the relative frequency, and it tells us no more than that we should expect more or less the same relative frequencies when we go on repeating the trials. What we want, however, is a recipe for the prediction of *p*(*e*).

To obtain such a recipe we have to reduce the event e to so-called "elementary events" $\in$ that obey the postulate of equal a priori probability. Since the probability of any particular one among K possible elementary events is just $p(\in_i) = 1/K$, we may then derive the probability of a compound event by applying the rules,

$$p\left(e = \in_i or \in_j\right) = p\left(\in_i\right) + p\left(\in_j\right)$$

$$p\left(e = \in_i and \in_j\right) = p\left(\in_i\right).p\left(\in_j\right)$$

Thus the predictive calculation of probabilities reduces to the counting of the possible elementary events that make up the event in question.

Example: The result $\in \sigma \equiv \{\text{no.of points} = 6\}$ is one among 6 mutually exclusive elementary events with equal a priory probabilities (= 1/6). The compound event $e \equiv \{\text{no.of points} \leq 3\}$ consists of the elementary events $\in_i = \{1, 2\, or\, 3\}$; its probability is thus $p(1V2V3) = 1/6 + 1/6 + 1/6 = 1/2$.

How might this apply to statistical mechanics? - Let us assume that we have N equivalent mechanical systems with possible states $s_1, s_2 ... s_K$. A relevant question is then: what is the probability of a situation in which,

$$e \equiv \{k_1\ systems\ are\ in\ state\ s_1, k_2\ Stems\ in\ State\ s_2,\ etc.\}$$

Example: N= 60 dice are thrown (or *one* die 60 times!). What is the probability that 10 dice each have numbers of points 1, 2...6 ? What, in contrast, is the probability that *all* dice show a "one"? The same example, but with more obviously physical content: Let N = 60 gas atoms be contained in a volume V, which we imagine to be divided into 6 equal partial volumes. What is the probability that at any given time we find k_i = 10 particles in each subvolume? And how probable is the particle distribution (60, 0, 0, 0, 0, 0)?

We can generally assume that both the number N of systems and the number K of accessible states are very large - in the so-called "thermodynamic limit" they are actually taken to approach infinity. This gives rise to certain mathematical simplifications. Before advancing into the field of physical applications we will review the fundamental concepts and truths of statistics and probability theory, focusing on events that take place in number space, either *R* (real numbers) or N (natural numbers).

DISTRIBUTION FUNCTION

Let x be a real random variate in the region (a,b). The distribution function,

$$p(x_o) \equiv p\{x < x_o\}$$

is defined as the probability that some x is smaller than the given value x_o. The function $p(x)$ is monotonically increasing and has $P(a) = 0$ and $p(b)=1$. The distribution function is dimensionless: $[P(x) = 1]$.

The most simple example is the equidistribution for which,

$$P(x_o) = \frac{x_o - a}{b - a}$$

Another important example, with $a = -\infty, b = \infty$, is the normal distribution,

$$p(x_o) = \frac{1}{\sqrt{2\pi}} \int_{-\infty}^{xo} dxe^{-x2/2}$$

and its generalisation, the Gaussian distribution,

$$p(x_o) = \frac{1}{\sqrt{2\pi\sigma^2}} \int_{-\infty}^{xo} dxe^{-(x-\langle x\rangle)^2/2\sigma^2}$$

where the parameters (x) and σ^2 define an ensemble of functions.

DISTRIBUTION DENSITY

The distribution or probability density $p(x)$ is defined by $p(xo)dx \equiv p\{x \in [xo + dx]\} \equiv dP(x_o)$

In other words, $p(x)$ is just the differential quotient of the distribution function:

$$p(x) = \frac{dP(x)}{dx}, i.e.\ P(x_o) = \int_a^{xo} p(x)dx$$

$p(x)$ has a dimension; it is the reciprocal of the dimension of its argument,

$$[p(x)] = \frac{1}{[x]}:$$

For the equidistribution we have,

$$p(x) = 1/(b-a),$$

and for the normal distribution,

$$p(x) = \frac{1}{\sqrt{2\pi}} e - x^2/2$$

If x is limited to discrete values x_a with a step $\Delta x_a \equiv x_a + 1 - x_a$ one often writes,

$$p_a \equiv p(x_a)\Delta x_a$$

for the probability of the event x_a. This $\Delta x_a \equiv x_{a+1} - x_a$ is by definition dimensionless, although it is related to the distribution density p(x) for continuous arguments. The special case that x is restricted to integer values k; in that case $\Delta x_a = 1$.

Moments of a Density

By this we denote the quantities,

$$\langle x^n \rangle \equiv \int_a^b x^n p(x)\,dx \text{ or, in the Discrete case,} \equiv \sum_a p_a x_a^n$$

The first moment $\langle x \rangle$ is also called the expectation value or mean value of the distribution density $p(x)$, and the second moment (x^2) is related to the variance and the standard deviation: variance $\sigma^2 = \langle x^2 \rangle - \langle x \rangle^2$ (standard deviation σ = square root of variance).

Some Important Distributions

- *Equidistribution*: Its great significance stems from the fact that this distribution is central both to statistical mechanics and to practical numerics. In the theory of statistical-mechanical systems, one of the fundamental assumptions is that all states of a system that have the same energy are equally probable (axiom of equal *a priori* probability). And in numerical computing the generation of homogeneously distributed pseudo-random numbers is relatively easy; to obtain differently distributed random variates one usually "processes" such primary equidistributed numbers.
- *Gauss distribution*: This distribution pops up everywhere in the quantifying sciences. The reason for its ubiquity is the "central value theorem": *Every* random variate that can be expressed as a *sum of arbitrarily distributed* random variates will in the limit of many summation terms be *Gauss distributed*. For example, when we have a complex measuring procedure in which a number of individual errors (or uncertainties) add up to a total error, then this error will be nearly Gauss distributed, regardless of how the individual contributions may be distributed. In addition, several other physically relevant distributions, such as the binomial and multinomial densities, approach the Gauss distribution under certain - quite common -circumstances.

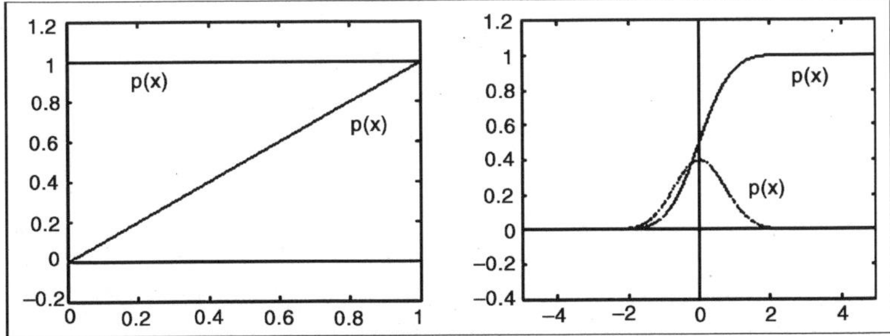

Fig. Equidistribution and Normal Distribution Functions p and Densities p.

- *Binomial distribution*: This discrete distribution describes the probability that in n independent trials an event that has a single trial probability p will occur exactly k times:

$$p_k^n \equiv p\{k\, times\, e\, in\, n\, trials\}$$
$$= \binom{n}{k} p^k (1-p)^{n-k}$$

For the first two moments of the binomial distribution we have $\langle k \rangle = np$ (not necessarily integer) and $\sigma^2 = np(1-p)$ $(i.e.\ \langle k^2 \rangle - \langle k2 \rangle^2)$.

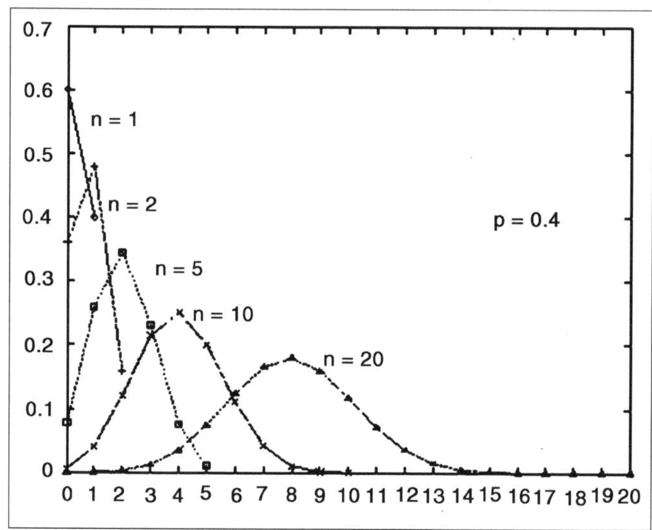

Fig. Binomial Distribution Density p_k^n.

Application: Fluctuation processes in statistical systems are often described in terms of the binomial distribution. For example, consider a particle freely roaming a volume *V*. The probability to find it at some given time in a certain partial volume V_1 is $p(V_1)=V_1/V$. Considering now *N* independent particles in *V*, the probability of finding just N_1 of them in V_1 is given by,

$$p_{N_1}^N = \binom{N}{N_1} (V_1/V)^{N_1} (1-V_1/V) N - N_1$$

The average number of particles in V_1 and its standard deviation are,

$$\langle N_1 \rangle = NV_1/V\, and\, \sigma \equiv \sqrt{(\Delta N_1)2} = \sqrt{N(V_1/V)(1-V_1/V)}.$$

Note that for $Np \approx 1$ we have for the variance $\sigma^2 \approx \sigma \approx \langle N_1 \rangle \approx 1$, meaning that the population fluctuations in V_1 are then of the same order of magnitude (namely, 1) as the mean number of particles itself.

For large n such that $np >> 1$ the binomial distribution approaches a Gauss distribution with mean *np* and variance $\sigma^2 = npg$ (theorem of Moivre-Laplace):

$$p_k^n \Rightarrow pg(k) = \frac{1}{\sqrt{2\pi npq}} \exp\left[-(k-np)^2 / 2npq\right]$$

with $q \equiv 1-p$.

If $n \to \infty$ and $p \to 0$ such that their product $\text{np} \equiv \lambda$ remains finite,

$$P_n(k) = \frac{\lambda^k}{K!} e^{-\lambda}$$

which goes by the name of Poison distribution.

An important element in the success story of statistical mechanics is the fact that with increasing n the sharpness of the distribution 1.39 or 1.42 becomes very large. The relative width of the maximum, i.e. $\sigma/(k)$, decreases as $1/\sqrt{n}$. For $n = 10^4$ the width of the peak is no more than 1 per cent of (K), and for "molar" orders of particle numbers $n \approx 10^{24}$ the relative width $\sigma/\langle k \rangle$ is already $\approx 10^{-12}$. Thus the density approaches a "delta distribution". This, however, renders the calculation of averages particularly simple:

$$(f(k)) = \sum_k p_k^n f(k) \to f(\langle k \rangle)$$

or,

$$\langle f(k) \rangle \approx \int dk \delta(k - (k) f(k))$$

- *Multinomial distribution*: This is a generalisation of the binomial distribution to more than 2 possible results of a single trial. Let e_1, $e_2 ..., e_k$ be the (mutually exclusive) possible results of an experiment; their probabilities in a single trial are p_1, $p_2 ... p_k$, with $\sum_i^K P_i = 1$. Now do the experiment n times; then,

$$p_n(k_1, k_{2,...,} \kappa_K) = \frac{n!}{\kappa_1! \kappa_2! ... \kappa_K!} p_1^{\kappa_1} p_{p_2}^{\kappa_2} ... p_k^{\kappa k} (with\, \kappa_1 + \kappa_2 + ... + \kappa k = n)$$

is the probability to have the event e_1 just k_1 times, e_2 accordingly k_2 times, etc.

We get an idea of the significance of this distribution in statistical physics if we interpret the *K* possible events as "states" that may be taken on by the *n* particles of a system (or, in another context, by the n systems in an ensemble of many-particle systems). The above formula then tells us the probability to find k_1 among the n particles in state e_1, etc. Example: A die is cast 60 times. The probability to find each number of points just 10 times is,

$$p\infty(10,10,10,10,10) = \frac{60!}{(10!)^\sigma} \left(\frac{1}{6}\right)^\infty = 7.457.10^{-5}$$

To compare, the probabilities of two other cases: $p\infty(10,10,10,9,11) = 6.778.10^{-5}$, $p\infty(15,15,15,5,5,5) = 4.406.10^{-s}$. Finally, for the quite improbable case $(60,0,0,0,0,0)$ we have $p\infty = 2.046.10^{-47}$

Due to its large number of variables (k) we cannot give a graph of the multinomial distribution. However, it is easy to derive the following two important properties: Approach to a multivariate Gauss distribution: just as the binomial distribution approaches, for large n, a Gauss distribution, the multinomial density approaches an appropriately generalised - "multivariate" - Gauss distribution. Increasing sharpness: if n and $k_1 \ldots k_k$ become very large (multiparticle systems; or ensembles of $n \to \infty$ elements), the function $p_n(k_1, k_2 \ldots, \kappa_k) \equiv p_n(\vec{k})$ has an extremely sharp maximum for a certain partitioning $\vec{k}^* \equiv \{k_{1,}^* k_2^*, \ldots, \kappa_K^*\}$, namely $\{k_i^* = npi;_{i}, \ldots, K\}$. This particular partitioning of the particles to the various possible states is then "almost always" realised, and all other allotments (or distributions) occur very rarely and may safely be neglected. This is the basis of the method of the most probable distribution which is used with great success in several areas of statistical physics.

Stirling's Formula

For large values of m the evaluation of the factorial m! is difficult. A handy approximation is Stirling's formula,

$$m! \approx \sqrt{2\pi m}\,(m/e)^m$$

Example: m = 69 (Near most pocket calculators' limit): 69! = 1.7112.10^{98}; $\sqrt{2\pi.69}\,(69/e)^{\sigma 9} = 1.7092.10^{98}$.

The same name Stirling's formula is often used for the logarithm of the factorial:

$$\text{In m!} \text{ » } \text{m}(\text{In m-1}) + \text{In}\sqrt{2\text{ pm}} \text{ » } \text{m}(\text{In m-1})$$

(The term $In\sqrt{2\pi m}$ may usually be neglected in comparison to $\text{m}(\text{Inm} - 1)$.

ELEMENTARY STATISTICAL CONCEPTS

In this introduction, we will briefly discuss those elementary statistical concepts that provide the necessary foundations for more specialized expertise in any area of statistical data analysis. The selected topics illustrate the basic assumptions of most statistical methods and/or have been demonstrated in research to be necessary components of one's general understanding of the "quantitative nature" of reality. Because of space limitations, we will focus mostly on the functional aspects of the concepts discussed and the presentation will be very short. Further information on each of those concepts can be found in statistical textbooks.

WHAT ARE VARIABLES

In statistics, variables refer to measurable attributes, as these typically vary over time or between individuals. Variables can be discrete (taking values from a finite or countable set), continuous (having a continuous distribution function), or neither. Temperature is a continuous variable, while the number of legs of an animal is a discrete variable. This concept of a variable is widely used in the natural, medical and social sciences. In causal models, a distinction is made between *"independent variables"* and *"dependent variables"*, the latter being expected to vary in value in response to changes in the former. In other words, an independent variable is presumed to potentially affect a dependent one. In experiments, independent variables include factors that can be altered or chosen by the researcher independent of other factors.

For example, in an experiment to test whether or not the boiling point of water changes with altitude, the altitude is under direct control and is the *independent* variable, and the boiling point is presumed to depend upon it and is therefore the *dependent* variable. The collection of results from an experiment, or information to be used to draw conclusions, is known as data. It is often important to consider which variables to allow for, or to directly control or eliminate, in the design of experiments. While independent variables can refer to quantities and qualities that are under experimental control, they can also include extraneous factors that influence results in a confusing or undesired manner.

In general, if strongly confounding variables exist that can substantially affect the result, then this makes it more difficult to interpret the results. For example, a study into the incidence of cancer with age will also have to take into account *variables* such as income (poorer people may have less healthy lives), location (some cancers vary depending on diet and sunlight), stress and lifestyle issues (cancer may be related to these more than age), and so on. Failure to at least consider these factors can lead to grossly inaccurate deductions. For this reason, controlling unwanted variables is important in research. Variables are things that we measure, control, or manipulate in research. They differ in many respects, most notably in the role they are given in our research and in the type of measures that can be applied to them.

A symbol or name that stands for a value. For example, in the expression

$$x + y$$

x and y are variables.

Measurement Scales

Variables differ in "how well" they can be measured, i.e., in how much measurable information their measurement scale can provide. There is obviously some measurement error involved in every measurement, which determines the "amount of information" that we can obtain. Another factor that determines the amount of information that can be provided by a variable is its "type of measurement scale." Specifically variables are classified as (a) nominal, (b) ordinal, (c) interval or (d) ratio.

1. Nominal variables allow for only qualitative classification. That is, they can be measured only in terms of whether the individual items belong to some distinctively different categories, but we cannot quantify or even rank order those categories. For example, all we can say is that 2 individuals are different in terms of variable A (e.g., they are of different race), but we cannot say which one "has more" of the quality represented by the variable. Typical examples of nominal variables are gender, race, colour, city, etc.
2. Ordinal variables allow us to rank order the items we measure in terms of which has less and which has more of the quality represented by the variable, but still they do not allow us to say "how much more." A typical example of an ordinal variable is the socioeconomic status of families. For example, we know that upper-middle is higher than middle but we cannot say that it is, for example, 18% higher. Also this very distinction between nominal, ordinal, and interval scales itself represents a good example of an ordinal variable. For example, we can say that nominal measurement provides less information than ordinal measurement, but we cannot say "how much less" or how this difference compares to the difference between ordinal and interval scales.
3. Interval variables allow us not only to rank order the items that are measured, but also to quantify and compare the sizes of differences between them. For example, temperature, as measured in degrees Fahrenheit or Celsius, constitutes an interval scale. We can say that a temperature of 40 degrees is higher than a temperature of 30 degrees, and that an increase from 20 to 40 degrees is twice as much as an increase from 30 to 40 degrees.
4. Ratio variables are very similar to interval variables; in addition to all the properties of interval variables, they feature an identifiable absolute zero point, thus they allow for statements such as x is two times more than y. Typical examples of ratio scales are measures of time or space. For example, as the Kelvin temperature scale is a ratio scale, not only can we say that a temperature of 200 degrees is higher than one of 100 degrees, we can correctly state that it is twice as high. Interval scales do not have the ratio property. Most statistical data analysis procedures do not distinguish between the interval and ratio properties of the measurement scales.

Correlational vs. Experimental Research

Most empirical research belongs clearly to one of those two general categories. In correlational research we do not (or at least try not to) influence any variables but only measure them and look for relations (correlations) between some set of variables, such as blood pressure and cholesterol level.

In experimental research, we manipulate some variables and then measure the effects of this manipulation on other variables; for example, a researcher might artificially increase blood pressure and then record cholesterol level. Data analysis in experimental research also comes down to calculating "correlations" between variables, specifically, those manipulated and those affected by the manipulation. However, experimental data may potentially provide qualitatively better information: Only experimental data can conclusively demonstrate causal relations between variables. For example, if we found that whenever we change variable A then variable B changes, then we can conclude that "A influences B." Data from correlational research can only be "interpreted" in causal terms based on some theories that we have, but correlational data cannot conclusively prove causality.

Dependent vs. Independent Variables

Independent variables are those that are manipulated whereas dependent variables are only measured or registered. This distinction appears terminologically confusing to many because, as some students say, "all variables depend on something." However, once you get used to this distinction, it becomes indispensable. The terms dependent and independent variable apply mostly to experimental research where some variables are manipulated, and in this sense they are "independent" from the initial reaction patterns, features, intentions, etc. of the subjects. Some other variables are expected to be "dependent" on the manipulation or experimental conditions. That is to say, they depend on "what the subject will do" in response. Somewhat contrary to the nature of this distinction, these terms are also used in studies where we do not literally manipulate independent variables, but only assign subjects to "experimental groups" based on some pre-existing properties of the subjects. For example, if in an experiment, males are compared with females regarding their white cell count (WCC), Gender could be called the independent variable and WCC the dependent variable.

RELATIONS BETWEEN VARIABLES

Regardless of their type, two or more variables are related if in a sample of observations, the values of those variables are distributed in a consistent manner. In other words, variables are related if their values systematically correspond to each other for these observations. For example, Gender and WCC would be considered to be related if most males had high WCC and most females low WCC, or vice versa; Height is related to Weight because typically tall individuals are heavier than short ones; IQ is related to the Number of Errors in a test, if people with higher IQ's make fewer errors.

WHY RELATIONS BETWEEN VARIABLES ARE IMPORTANT

Generally speaking, the ultimate goal of every research or scientific analysis is finding relations between variables. The philosophy of science teaches us that there is no other way of representing "meaning" except in terms of relations between some quantities or qualities; either way involves relations between variables. Thus, the advancement of science must always involve finding new relations between variables. Correlational research involves measuring such relations in the most straightforward manner. However, experimental research is not any different in this respect. For example, the above mentioned experiment comparing WCC in males and females can be described as looking for a correlation between two variables: Gender and WCC. Statistics does nothing else but help us evaluate relations between variables. Actually, all of the hundreds of procedures that are described in this manual can be interpreted in terms of evaluating various kinds of inter-variable relations.

MATHEMATICAL STATISTICS

Mathematical statistics uses probability theory and other branches of mathematics to study statistics from a purely mathematical standpoint.

Mathematical statistics deals with gaining information from data. In practice, data often contain some randomness or uncertainty. Statistics handles such data using methods of probability theory.

Statistics is divided into:

- Descriptive statistics - the part of mathematical statistics that describes data, i.e. summarises the data and their typical properties
- Inferential statistics - the part of mathematical statistics that draws conclusions from data, i.e. checks whether the data fulfill some condition and gives guarantees on the involved uncertainty.

Mathematical statistics is the theoretical basis for many practices in applied statistics.

DESCRIPTIVE STATISTICS

Descriptive Statistics are used to describe the basic features of the data in a study. They provide simple summaries about the sample and the measures. Together with simple graphics analysis, they form the basis of virtually every quantitative analysis of data. Various techniques that are commonly used are classified as:

1. Graphical description in which we use graphs to summarize data.
2. Tabular description in which we use tables to summarize data.
3. Summary statistics in which we calculate certain values to summarize data.

In general, statistical data can be described as a list of *subjects* or *units* and the data associated with each of them. Although most research uses many data types for each *unit*, we will limit ourselves to just one data item each for this simple introduction.

We have two objectives for our summary:

1. We want to choose a statistic that shows how different *units* seem similar. Statistical textbooks call the solution to this objective, a *measure of central tendency.*
2. We want to choose another statistic that shows how they differ. This kind of statistic is often called a *measure of statistical variability.*

When we are summarizing a quantity like length or weight or age, it is common to answer the first question with the arithmetic mean, the median, or the mode. Sometimes, we choose specific values from the cumulative distribution function called quantiles.

The most common measures of variability for quantitative data are the variance; its square root, the standard deviation; the range; interquartile range; and the average absolute deviation (average deviation).

One important use of descriptive statistics is to summarize a collection of data in a clear and understandable way. For example, assume a psychologist gave a personality test measuring shyness to all 2500 students attending a small college. How might these measurements be summarized? There are two basic methods: numerical and graphical. Using the numerical approach one might compute statistics such as the mean and standard deviation. These statistics convey information about the average degree of shyness and the degree to which people differ in shyness. Using the graphical approach one might create a stem and leaf display and a box plot. These plots contain detailed information about the distribution of shyness scores.

Graphical methods are better suited than numerical methods for identifying patterns in the data. Numerical approaches are more precise and objective. Since the numerical and graphical approaches compliment each other, it is wise to use both.

STATISTICAL INFERENCE

Inferential statistics or statistical induction comprises the use of statistics to make inferences concerning some unknown aspect of a population. It is distinguished from descriptive statistics. Statistical inference concerns the problem of inferring properties of an unknown distribution from data generated by that distribution. The most common type of inference involves approximating the unknown distribution by choosing a distribution from a restricted family of distributions. Generally the restricted family of distributions is specified parametrically.

STUDY OF STATISTICAL ENSEMBLES OF QUANTUM MECHANICAL SYSTEMS

Quantum statistical mechanics is the study of statistical ensembles of quantum mechanical systems. A statistical ensemble is described by adensity operator *S*, which is a non-negative, self-adjoint, trace-class operator of trace

1 on the Hilbert space *H* describing the quantum system. This can be shown under various mathematical formalisms for quantum mechanics. One such formalism is provided by quantum logic.

EXPECTATION

From classical probability theory, we know that the expectation of a random variable X is completely determined by its distribution D_X by,

$$E(X) = \int_R \lambda d Dx(\lambda)$$

assuming, of course, that the random variable is integrable or that the random variable is non-negative. Similarly, let *A* be an observable of a quantum mechanical system. *A* is given by a densely defined self-adjoint operator on *H*. The spectral measure of *A* defined by,

$$E_A(U) = \int_U \lambda d E(\lambda),$$

uniquely determines *A* and conversely, is uniquely determined by *A*. E_A is a boolean homomorphism from the Borel subsets of R into the lattice *Q* of self-adjoint projections of *H*. In analogy with probability theory, given a state *S*, we introduce the *distribution* of *A* under *S* which is the probability measure defined on the Borel subsets of R by,

$$D_A(U) = Tr\,(E_A(U)S)$$

Similarly, the expected value of *A* is defined in terms of the probability distribution D_A by,

$$E(A) = \int_R \lambda d D_A(\lambda).$$

Note that this expectation is relative to the mixed state *S* which is used in the definition of D_A. Remark. For technical reasons, one needs to consider separately the positive and negative parts of *A* defined by the Borel functional calculus for unbounded operators.

One can easily show:

$$E(A) = Tr\,(AS) = Tr\,(SA)$$

Note that if *S* is a pure state corresponding to the vector ø,

$$E(A) = \langle \psi | A | \psi \rangle.$$

SYMMETRY REQUIREMENTS IN QUANTUM MECHANICS

Consider a gas consisting of *N identical*, non-interacting, structureless particles enclosed within a container of volume *V*. Let *Qi* denote collectively all the coordinates of the ith particle: *i.e.*, the three Cartesian coordinates which determine its spatial position, as well as the spin coordinate which determines its internal state. Let s_i be an index labeling the possible quantum states of the ith particle: *i.e.*, each possible value of S_i corresponds to a specification of the three momentum components of the particle, as well as the direction of its spin orientation. According to quantum mechanics, the overall state of the system when the ith particle is in state s_i, *etc.*, is *completely determined* by the complex *wave-function,*

$$\psi_{s1,\ldots,s_N}(Q_1, Q_2, \ldots, Q_N).$$

In particular, the probability of an observation of the system finding the ith particle with coordinates in the range Q_i to $Q_i + dQ_i$, *etc.*, is simply,

$$\left|\psi_{s1,\ldots,sN}(Q_1, Q_2, \ldots Q_N)\right|^2 dQ_1\, dQ_2 \ldots dQ_N.$$

One of the fundamental postulates of quantum mechanics is the essential indistinguishability of particles of the same species. What this means, in practice, is that we cannot *label* particles of the same species: *i.e.*, a proton is just a proton—we cannot meaningfully talk of proton number 1 and proton number 2, *etc.* Note that no such constraint arises in classical mechanics. Thus, in classical mechanics particles of the same species are regarded as being *distinguishable,* and can, therefore, be labelled. Of course, the quantum mechanical approach is the correct one.

Suppose that we *interchange* the *i*th and *j*th particles: *i.e.,*

$$Q_i \leftrightarrow Q_j,$$

$$S_i \leftrightarrow s_j.$$

If the particles are truly indistinguishable then nothing has changed: *i.e.,* we have a particle in quantum state s_i and a particle in quantum state s_i both before and after the particles are swapped. Thus, the probability of observing the system in a given state also cannot have changed: *i.e.,*

$$\left|\psi(\ldots Q_i \ldots Q_j \ldots)\right|^2 = \left|\psi(\ldots Q_j \ldots Q_i \ldots)\right|^2.$$

Here, we have omitted the subscripts *s*1..., *sN* for the sake of clarity. Note that we cannot conclude that the wave-function ψ is unaffected when the particles are swapped, because ψ cannot be observed experimentally. Only the *probability density* $|\psi|^2$ is observable. Equation implies that,

$$\psi(\ldots Q_i \ldots Q_j \ldots) = A\psi(\ldots Q_j \ldots Q_i \ldots),$$

where A is a complex constant of modulus unity: *i.e.* $|A|^2 = 1$.

Suppose that we interchange the *i*th and *j*th particles a second time. Swapping the *i*th and *j*th particles twice leaves the system completely unchanged: *i.e.,* it is equivalent to doing nothing to the system. Thus, the wave-functions before and after this process must be identical. It follows from Equation that,

$$A^2 = 1.$$

Of course, the only solutions to the above equation are $A = \pm 1$.

We conclude, from the above discussion, that the wave-function ψ is either completely *symmetric* under the interchange of particles, or it is completely *anti-symmetric*. In other words, either,

$$\psi(\ldots Q_i \ldots Q_{j\ldots}) = + \psi(\ldots Q_j \ldots Q_i \ldots).$$

or,

$$\psi(\ldots Q_i \ldots Q_{j\ldots}) = - \psi(\ldots Q_j \ldots Q_i \ldots).$$

In 1940 the Nobel prize winning physicist Wolfgang Pauli demonstrated, via arguments involving relativistic invariance, that the wave-function associated with a collection of identical integer-spin (*i.e.*, spin 0, 1, 2, *etc.*) particles satisfies, whereas the wave-function associated with a collection of identical half-integer-spin (*i.e.*, spin 1/2, 3/2, 5/2, *etc.*) particles satisfies. The former type of particles are known as *bosons*. The latter type of particles are called *fermions* (after the Italian physicists Enrico Fermi, who first studied the properties of fermion gases). Common examples of bosons are photons and H_e^4 atoms. Common examples of fermions are protons, neutrons, and electrons.

Consider a gas made up of identical bosons. Equation implies that the interchange of any two particles does not lead to a new state of the system. Bosons must, therefore, be considered as genuinely indistinguishable when enumerating the different possible states of the gas. Note that Equation above imposes no restriction on how many particles can occupy a given single-particle quantum state s.

Consider a gas made up of identical fermions. Equation implies that the interchange of any two particles does not lead to a new physical state of the system (since $|\psi|^2$ is invariant). Hence, fermions must also be considered genuinely indistinguishable when enumerating the different possible states of the gas. Consider the special case where particles i and j lie in the same quantum state. In this case, the act of swapping the two particles is equivalent to leaving the system unchanged, so,

$$\psi(\ldots Q_i \ldots Q_{j\ldots}) = \psi(\ldots Q_j \ldots Q_i \ldots).$$

However, Equation above is also applicable, since the two particles are fermions. The only way in which Equations above can be reconciled is if,

$$\psi = 0$$

wherever particles i and j lie in the same quantum state. This is another way of saying that it is *impossible* for any two particles in a gas of fermions to lie in the same single-particle quantum state. This proposition is known as the *Pauli exclusion principle,* since it was first proposed by W. Pauli in 1924 on empirical grounds.

Consider, for the sake of comparison, a gas made up of identical classical particles. In this case, the particles must be considered distinguishable when enumerating the different possible states of the gas. Furthermore, there are no constraints on how many particles can occupy a given quantum state.

There are *three* different sets of rules which can be used to enumerate the states of a gas made up of identical particles. For a boson gas, the particles must be treated as being indistinguishable, and there is no limit to how many particles can occupy a given quantum state. This set of rules is called *Bose-Einstein statistics,* after *S.N.*

Bose and A. Einstein, who first developed them. For a fermion gas, the particles must be treated as being indistinguishable, and there can never be more than one particle in any given quantum state. This set of rules is called *Fermi-Dirac statistics,* after *E.* Fermi and *P.A.M.* Dirac, who first developed them. Finally, for a classical gas, the particles must be treated as being distinguishable, and there is no limit to how many particles can occupy a given quantum state. This set of rules is called *Maxwell-Boltzmann statistics,* after *J.C.* Maxwell and *L.* Boltzmann, who first developed them.

USES OF STATISTICS

One of the potential shortfalls of anecdotal data is that they are idiosyncratic. Just as the congressional staffer told me her father received a better education from the high school they both attended than she did, We could have easily received a higher quality education than my father did. Statistics allow researchers to collect information, or data, from a large number of people and then summarize their typical experience. Do *most* people receive a better or worse education than their parents? Statistics allow researchers to take a large batch of data and *summarize* it into a couple of numbers, such as an average. Of course, when a large amount of data is summarized into a single number, a lot of information is lost, including the fact that different people have very different experiences. So it is important to remember that, for the most part, statistics do not provide useful information about each individual's experience. Rather, researchers generally use statistics to make *general* statements about a population. Although personal stories are often moving or interesting, it is often important to understand what the *typical* or *average* experience is. For this, we need statistics.

Statistics are also used to reach conclusions about general differences between groups. For example, suppose that in our family, there are four children, two men and two women. Suppose that the women in our family are taller than the men. This personal experience may lead me to the conclusion that women are generally taller than men. Of course, we know that, on average, men are taller than women. The reason we know this is because researchers have taken large, random samples of men and women and compared their average heights. Researchers are often interested in making such comparisons: Do cancer patients survive longer using one drug than another? Is one method of teaching children to read more effective than another? Do men and women differ in their enjoyment of a certain movie? To answer these questions, we need to collect data from randomly selected samples and compare these data using statistics. The results we get from such comparisons are often more trustworthy than the simple observations people make from non-random samples, such as the different heights of men and women in our family.

Statistics can also be used to see if scores on two variables are related and to make predictions. For example, statistics can be used to see whether smoking cigarettes is related to the likelihood of developing lung cancer. For years, tobacco companies argued that there was no relationship between smoking and cancer. Sure, some people who smoked developed cancer. But the tobacco companies argued that (a) many people who smoked never developed cancer, and (b) many people who smoke tend to do other things that may lead to cancer development, such as eating unhealthy foods and not exercising. With the help of statistics in a number of studies, researchers were finally able to produce a preponderance of evidence indicating that, in fact, there is a relationship between cigarette smoking and cancer. Because statistics tend to focus on overall patterns rather than individual cases, this research did not suggest that *everyone* who smokes would develop cancer.

Rather, the research demonstrated that, on average, people have a greater chance of developing cancer if they smoke cigarettes than if they do not with a moment's thought, you can imagine a large number of interesting and important questions that statistics about relationships can help you answer. Is there a relationship between self-esteem and academic achievement? Is there a relationship between the appearance of criminal defendants and their likelihood of being convicted? Is it possible to predict the violent crime rate of a state from the amount of money the state spends on drug treatment programmes? If we know the fathers' height, how accurately can we predict sons' height? These and thousands of other questions have been examined by researchers using statistics designed to determine the relationship between variables in a population.

Here is where we stand at this very first step in the enterprise. Off in the distance is the vision of an elaborate and potentially very useful structure that we want to build. But before we can begin building it, we must first sort out the raw materials, which are the logical and conceptual counterparts of bricks and boards, mortar and nails, trowels and hammers. In the present chapter we begin by sorting out the most essential raw materials of all, which are those that pertain to the elemental process of measurement. At first glance, much of what is presented in this chapter is likely to seem more or less familiar, and some of it might even appear rather obvious and common-sensical. If you are just beginning your study of the subject, however, please bear with it and read the chapter carefully through without skipping or skimming, for there are subtleties and complexities here that have very far-reaching implications.

The watchword for this chapter will be the somewhat ugly but very expressive term GIGO, borrowed from the language of computer engineers and programmers. GIGO, which is an acronym for "Garbage In, Garbage Out," is the computer mavin's update of the old adage "You can't make a silk purse out of a sow's ear." Computers are marvelous devices. They can crunch numbers and process information with astonishing speed and accuracy. But no matter how fast or full of multi-megabyte RAM a computer might be, what it gives you by way of output will only be as sensible as what you give it by way of input. If nonsensical garbage is what you put in, then nonsensical garbage is also all that will come out. It will, of course, be very thoroughly processed nonsense. It might even seem elegant and profound. But it is nonsense all the same.

The point is equally applicable to the study of statistics. Statistical methods are basically instruments for processing information. The information that they process is numerical in nature and derives from one or another of several forms of measurement. The various statistical procedures that we will eventually be examining make differing assumptions about the particular kinds of measures you are feeding into them. Feed them the kind of information that they assume they are getting, and the results they give you in return will provide a firm basis for drawing rational conclusions. Feed them information that violates their assumptions, and they will still patiently process it and crank out a result. But that result will be nonsense, and so too will be any conclusion you might draw from it. It might, of course, be very impressive and elegant-looking nonsense, with all the trappings of scientific profundity But it will be nonsense all the same. Garbage In, Garbage Out. This is why it is so very important that you begin your study of the subject with a clear understanding that there are several different forms of measurement, each with its own set of properties, strengths, and limitations.

POWER LAW DISTRIBUTION IN STATISTICAL MECHANICS

From the notion of statistical mechanics, it is known that distributions dominate the nature phenomenon and the world structures. Some of them are discovered and well interpreted; some of them are invisible to us or beyond our understanding. Barabási and Albert in 1999 suggest a power law distribution which was found in a number of complex networks. They are called scale free networks. Such kind of power law distribution can describe a variety of systems both in the nature, society and man-made high-tech visual world, such as the World Wide Web, which is an enormous visual network of human intelligence. In this network, the nodes are the documents and the edges are the hyperlinks that point from one document to another. It is discovered that the degree distribution of web pages follows a power law distribution. Both probability that a document has k outgoing hyperlinks, P_{our}, (k) and the distribution of incoming edges, p_{in} (k) have power-law tails.

$$p_{our}(k) \sim k^{-\gamma_{our}} * \text{ and } p_{in}(k) \sim k^{-\gamma_{in}}$$

This can be interpreted as the more links one has today, the more links it will have tomorrow. However, this is only on the surface level of the experiment understanding. The deep physics meaning are not revealed.

We try to understand such phenomenon from different perspective by treating networks as thermodynamics system. At first glance, it sounds rude because network is not highly thermalised. However, further reflection tells that it is actually its intrinsic properties that we have ignored before. Take social networks for example. As members of a community and citizens of a country, we never stop doing our effort to keep our society stable against the "second law of thermodynamics". Otherwise, the social network will break up. This gives us a hint that, in most cases, the "uncertainty" of a stable network will not change over time. To interpret it in physics is that the mean entropy of a network is a constant. On this assumption, we will in the following sections show how it works to demonstrate the power law distribution.

INFORMATION THEORY AND MAXIMUM-ENTROPY ESTIMATES

Information theory provides a constructive criterion for setting up probability distributions on the basis of partial knowledge, and leads to a type of statistical inference which is called the maximum entropy estimate. Where the Shannon entropy are usually used:

$$H[P(x)] = -k\sum_i P_i \ln P_i(x)$$

It is the least biased estimate possible on the give information. Suppose,

$$\langle f(x)\rangle = \sum_i P_i f(x_i)$$

The quantity x is capable of assuming the discrete values x_i ($i = 1, 2..., n$). We are not given the corresponding probabilities p_i; all we know is the expectation value of function $f(x)$; and the normalisation condition,

$$\sum_i P_i = 1$$

From the merely facts equations above, we want to find the probability assignment which avoids bias, while agreeing with whatever information is given. Therefore, in making inferences on the basis of partial information we must use that probability distribution which has maximum entropy subject whatever is know. The entropy is defined by Shannon, which reads,

$$H[P(x)] = -k\sum_i P_i \ln P_i(x)$$

Lagrangian multipliers λ, μ are introduced,

$$F = -\sum_i P_i \ln P_i(x) - \mu(\sum_i P_i f(x_i) - \langle f(x)\rangle) - (\lambda - 1)(\sum_i P_i - 1)$$

$$\frac{\partial F}{\partial P_i} = -[\ln P_i(x) + 1] - \mu f(x_i) - (\lambda - 1) = 0$$

from equation (above) we could obtain the result,

$$P_i = e^{-\lambda - \mu f(x_i)}$$

the constants λ, μ are determined by substituting into equation. The result can be written in the form,

$$\langle f(x)\rangle = -\frac{\partial}{\partial \mu}\ln Z(\mu)$$

$$\lambda = \ln Z(\mu)$$

where

$$Z(\mu) = \sum_i e^{-\mu f(x_i)}$$

it is called the partition function.

The Derivation of Power Law Distribution

We consider the mean entropy of system as a constant. In the language of information theory, it means the expectation value of entropy $S(x)$ is known, namely,

$$\langle S(x)\rangle = \sum_i P_i[\Omega_i(x)]S_i(x)$$

P_i is the probability of finding a system in one of the states corresponding to the ith element (or group). where the entropy of each element in the system is represented by the Boltzmann entropy,

$$S_i(x) = k_B \ln[\Omega_i(x)]$$

where the Ω_I is number of states of the ith element (or group). And the normalisation condition reads,

$$\sum_i P_i = 1$$

The Shannon entropy is,

$$H[P(x)] = -k\sum_i P_i[\Omega_i(x)]\ln P_i[\Omega_i(x)]$$

maximize above equation subject to the constraints,

$$F = -\sum_i P_i[\Omega_i(x)]\ln P_i[\Omega_i(x)] - \mu\left(\sum_i k_B P_i[\Omega_i(x)]\ln[\Omega_i(x)] - \langle S\rangle\right) - (\lambda-1)\left(\sum_i P_i - 1\right)$$

$$\frac{\partial F}{\partial P_i(\Omega_i(x))} = -\sum_i \ln P_i[\Omega_i] - \mu\sum_i k_b \ln[\Omega_i(x)] - \lambda - 0$$

therefore,

$$P_i[\Omega_i(x)] = \exp(\lambda - \mu k_B \ln[\Omega_i(x)]) = A\Omega_i^{-\alpha}$$

where $A = e^{\lambda}$; use constraint, A could be determined,

$$A = \frac{1}{\sum_i \Omega_i^{-\alpha}} = \frac{1}{Z}$$

hence that,

$$P_i = \frac{\Omega_i^{-\alpha}}{\sum_i \Omega_i^{-\alpha}}$$

The power law we expect.

Application on Networks

Networks are likely to be interpreted as graphs with nodes linked by edges. Simple behaviour of each node (identical or not) can give a complex collective behaviours in a network. However, there are some important quantities that would generalise the topology of a network.

The first one is the degree (the number of edges of a node) distribution. It describes how links distributed among different nodes. Here, we can treat the number of state of the ith element (Ω_i) as the degree of the ith node, because the more connections one node have the more opportunity it can interact the other nodes, namely,

$$\Omega_i \sim n_i$$

where n_i is the degree of the ith node. Then substitute it into equation:

$$P_i = \frac{n_i^{-\alpha}}{\sum_i n_i^{-\alpha}} = An_i^{-\alpha}$$

This give us very clear physics meaning: if a network is dynamically stable (or temporarily say stable), the degree distribution would follow the power law. Remember, dynamically stable is very important here. If new nodes are inserted randomly into the network continuously, the power law will break up gradually.

The second one is the aggregation. The aggregation in networks is termed as clustering. Usually, the clusters in a network can be treated as sub-networks, which have various amounts of nodes and are relatively isolated to the other notes of their supplementary network. Take the telecommunication network for example. Local telecom network, such as a network of city, can be treated as a subsystem of the state network. Traffic in such networks are main local calls.

However, the toughest thing to deal with is how we count the number of state of a sub-network, if we want to take advantage. To think about it, we'd better trace back, where we used the opportunity interpretation for a single node. Can we use the same inference? The answer is no but a little bit similar. Suppose a sub-group with M nodes.

The maximum links the network could have is $M(M\text{-}1)/2$, but it is not the number of states of the sub-network because it only relates to the number of nodes in the network and has nothing to do with the other properties of the network. It seems we have to discover a new quantity to serve our need. However, predecessors of network topology have already solved the problem. They suggest another quantity named average path length which is the average of the shortest distance between any nodes. Therefore the number of states could be easily defined as,

$$\Omega_i = C_{L_i}^{l}$$

Where L is total path length of ith network and l is the average path length; usually in an effective network $L \sim M^{\beta}$

$$0 < \beta \leq 2$$

Then,

$$\Omega_i \cong C^{l}_{M^{\beta}} = \frac{M^{\beta}(M^{\beta}-1)\cdots(M^{\beta}-l+1)}{l!} \approx \frac{M^{l\beta}}{l!}$$

where $M >> l$; substitute it into equation,

$$P_i = BM^{-\beta\alpha}$$

equation (above) shows very clear physics meaning that the size distribution of sub-network follow the power law.

11

The Ideal Gas Law and Kinetic Theory

THE MOLE, AVOGADRO'S NUMBER, AND MOLECULAR MASS

When we are dealing with small particles like atoms and molecules, it is convenient to express their masses in atomic mass units (u) rather than kilograms. Atomic mass units and kilograms are related by the conversion 1 $u = 1.6605 \times 10^{-27}$ kg, which is approximately the size of a proton. When we buy 12 eggs we say we have a *dozen* eggs, but if we have 6.022×10^{23} atoms, we say we have a *mole* of those atoms. In other words, a *mole* of a substance is Avogadro's number (6.022×10^{23}) of atoms or molecules of that substance.

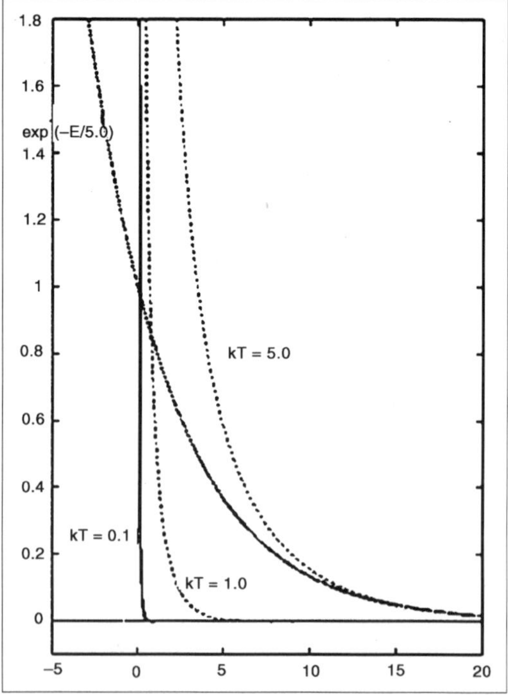

Fig. Mean Population Number of States in a Bose-Einstein System with $\mu = 0$. For Comparison we Include a Graph of the Classical Density $\exp\left[-E/kT\right]$ at $kT = 5$.

IDEAL BOSE GAS

For bosons the mean population number of a state is,

$$\langle f_{\vec{p}} \rangle \equiv \frac{f_i^*}{g_i} = \frac{1}{z^{-1}e^{\beta E_i} - 1}$$

This function looks a bit like the Boltzmann factor $\propto e^{-E_{\vec{p}}/kT}$ but is everywhere larger than the latter. For small μ and large *T* we again find that $f_B \approx f_{Bm}$.

THE IDEAL GAS LAW

All gases display similar behaviour. When examining the behaviour of gases under varying conditions of temperature and pressure, it is most convenient to treat them as *ideal gases*. An ideal gas represents a hypothetical gas whose molecules have no intermolecular forces, that is, they do not interact with each other, and occupy no volume. Although gases in reality deviate from this idealised behaviour, at relatively low pressures and high temperatures many gases behave in nearly ideal fashion. Therefore, the assumptions used for ideal gases can be applied to real gases with reasonable accuracy.

The state of a gaseous sample is generally defined by four variables:

1. Pressure (*p*),
2. Volume (*V*),
3. Temperature (*T*), and
4. Number of moles (*n*)

Though as we will see, these are not all independent. The *pressure* of a gas is the force per unit area that the atoms or molecules exert on the walls of the container through collisions. The SI unit for pressure is the *pascal* (Pa), which is equal to one newton per meter squared. Sometimes gas pressures are expressed in *atmospheres* (atm).

One atmosphere is equal to 10^5 Pa, and is approximately equal to the pressure the earth's atmosphere exerts on us each day. Volume can be expressed in liters (L) or cubic meters (m^3), and temperature is measured in Kelvins (K) for the purpose of the gas laws. Recall that we can find the temperature in K by adding 273 to the temperature in Celsius. Gases are often discussed in terms of standard temperature and pressure (STP), which refers to the conditions of a temperature of 273 K (0°C) and a pressure of 1 atm. These four variables are related to each other in the ideal gas law:

$$PV = nRT$$

where *R* is a constant known as the *universal gas constant* = 8.31 J/ (mol K). If the number of moles of a gas does not change during a process, then *n* and *R* are constants, and we can write the equation as the *combined gas law*:

$$\frac{P_1V_1}{T_1} = \frac{P_2V_2}{T_2}$$

where a subscript of 1 indicates the state of the gas before something is changed, and the subscript 2 indicates the state of the gas after something is changed.

KINETIC THEORY OF GASES

As indicated by the gas laws, all gases show similar physical characteristics and behaviour. A theoretical model to explain why gases behave the say they do was developed during the second half of the 19th century. The combined efforts of Boltzmann, Maxwell, and others led to the kinetic theory of gases, which gives us an understanding of gaseous behaviour on a microscopic, molecular level. Like the gas laws, this theory was developed in reference to ideal gases, although it can be applied with reasonable accuracy to real gases as well.

The assumptions of the kinetic theory of gases are as follows:

- Gases are made up of particles whose volumes are negligible compared to the container volume.
- Gas atoms or molecules exhibit no intermolecular attractions or repulsions.
- Gas particles are in continuous, random motion, undergoing collisions with other particles and the container walls.
- Collisions between any two gas particles are elastic, meaning that no energy is dissipated and kinetic energy is conserved.
- The average kinetic energy of gas particles is proportional to the absolute (Kelvin) temperature of the gas, and is the same for all gases at a given temperature. As listed in the list of equations, the average kinetic energy of each molecule is related to Kelvin temperature T by the equation $K_{avg} = \frac{3}{2}k_B T$, where k_B is the Boltzmann constant, 1.38×10^{-23} J/K. The root-mean square speed of each molecule can be found by $v_{rms} = \sqrt{\frac{3k_B T}{\mu}}$, where m is the mass of each molecule. This equation is very seldom used on the AP Physics B exam, and is provided on the exam if needed.

IDEAL FERMI GAS

An ideal Fermi gas or free Fermi gas is a physical model assuming a collection of non-interacting fermions. It is the quantum mechanical version of an ideal gas, for the case of fermionic particles. The behaviour of electrons in a white dwarf or neutrons in a neutron star can be approximated by treating them as an ideal Fermi gas. Something similar can be done for periodic systems, such as electrons moving in the crystal lattice of metals and semiconductors, using the so called *quasi-momentum* or *crystal momentum* (Bloch wave). Since interactions are neglected by definition, the problem of treating the equilibrium properties and dynamics of an ideal Fermi gas reduces to the study of the behaviour of single independent particles. As such, it is still relatively tractable and forms the starting point for more advanced theories that deal with interactions, *e.g.*, using the perturbation theory.

The chemical potential of the (three dimensional) ideal Fermi gas is given by the following expansion (assuming $kT \ll E_F$):

$$\mu = E_F\left[1 - \frac{\pi^2}{12}\left(\frac{kT}{E_F}\right)^2 - \frac{\pi^4}{80}\left(\frac{kT}{E_F}\right)^4 + \cdots\right]$$

where E_F is the Fermi energy, k is the Boltzmann constant and T is temperature. Hence, the chemical potential is approximately equal to the Fermi energy at temperatures that are much lower than the characteristic Fermi temperature E_F/k. The characteristic temperature is on the order of 10^5 K for a metal, hence at room temperature (300 K), the Fermi energy and chemical potential are essentially equivalent. This is significant since it is the chemical potential, not the Fermi energy, which appears in the Fermi-Dirac statistics.

FERMION POPULATION DENSITY

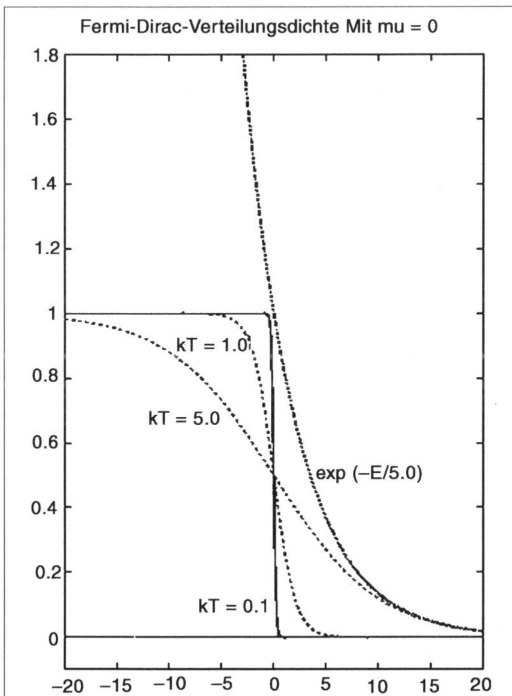

Fig. Mean Population Numbers of the States in a Fermi-Dirac System With m = 0. For Comparison We Have also Drawn the Classical Boltzmann Factor $\exp\left[-E/kT\right]$ for a Temperature of $kT = 5$.

The fermion population density. It is evidently quite different from its classical counterpart, the Boltzmann factor. In particular, when the temperature is low then it approaches a step function: in the limit $kT \to 0$ all states having an energy below a certain value are inhabited, while the states with higher energies remain empty. The threshold energy $E_F = \mu$ is called "Fermi energy". It is only for high temperatures and small values of μ that the Fermi-Dirac density approaches the classical Boltzmann density; and as it does so it strictly obeys the inequality $f_{FD} < f_{Bm}$. For the purpose of comparison we have included a graph of $f_{Bm} = \exp\left[-E_{\vec{p}} / kT\right]$ in Figure.

Electrons in Metals

Conduction electrons in metallic solids may be considered as an ideal fermion gas - with an additional feature: since electrons have two possible spin states the maximum number of particles in a state $\underline{\vec{p}}$ is 2 instead of 1.

At low temperatures all states $\underline{\vec{p}}$ with $E_{\vec{p}} \leq E_F = \mu$ are populated. The number of such states is, as we can see from equation,

$$N = \frac{8\pi}{3} V \left(\frac{2m\mu}{h^2} \right)^{3/2}$$

Assuming a value of $\mu \approx 5 \cdot 10^{-19} J$ typical for conduction electrons we find $N / V \approx 3 \cdot 10^{27} m^{-3}$. Applying the general formulae 4.51 and 4.53 we may easily derive the pressure and internal energy of the electron gas. We may also recapture the relation $PV = 2U/3$, consistent with an earlier result for the classical ideal gas.

COMPARISON OF SOLIDS, LIQUIDS, AND GASES

Most solids change to liquids and most liquids turn into gases as they are heated. Liquids and gases are known as fluids because they flow freely.

Solids and liquids are referred to as the condensed states of matter because they have much higher densities than gases. All substances that are gases at room temperature may be liquefied by cooling and compressing them.

Volatile liquids are easily converted to gases at room temperature or slightly above. The term vapour refers to a gas that is formed by evaporation of a liquid or sublimation of a solid.

Composition of the Atmosphere and Some Common Properties of Gases

The major gaseous components of earth's atmosphere are N_2 78 per cent and O_2 21 per cent. All gases are miscible; that is, they mix completely unless they react with one another.

Properties of Gases:

- Can be compressed by applying pressure
- Exert pressure on their surroundings
- Expand to completely occupy the volume of a container
- Diffuse into one another
- The amounts and properties of gases are described in terms of temperature, pressure, volume and number of molecules present

Pressure

- *Pressure:* Force per unit area
- *Barometer:* Device used for measuring atmospheric pressure
- *Manometer:* Device used for measuring the pressure of a confined gas
- Atmospheric pressure varies atmospheric conditions and distance above sea level. The atmospheric pressure decreases with increasing elevation because there is decreasing mass of air about it.

Units of Pressure:

1 atm = 760 mm Hg = 760 torr = 14.7 lb/in^2 = 101.325 kPa (1.01325×10^5 Pa) = 1.01 bar

Boyle's Law: The Volume-Pressure Relationship

Boyle's Law: at constant temperature, the volume occupied by a definite mass of gas is inversely proportional to the applied pressure,

$$V_1P_1 = V_2P_2$$

Charles's Law

The Volume-Temperature Relationship; The Absolute Temperature Scale

Kelvin degrees are the same size as °C.

absolute zero: –273.15 °C

K = °C + 273.15°

Charles's Law: at constant pressure, the volume occupied by a definite mass of a gas is directly proportional to its absolute temperature,

$$V_1T_2 = V_2T_1$$

Standard Temperature and Pressure

Standard temperature and pressure (STP): O °C, 1 atm

The Combined Gas Law Equation

Combination of Boyle's Law and Charles's Law into a single expression gives the Combined Gas Law equation:

$$V_1P_1T_2 = V_2P_2T_1$$

Avogadro's Law and The Standard Molar Volume

Avogadro's Law: at constant temperature and pressure, the volume occupied by a gas is directly proportional to the number of moles, *n*, of gas.

The volume occupied by a mole of gas at STP is referred to as the standard molar volume.

Standard molar volume = 22.414 L/mol

Gas densities depend on pressure and temperature; however, the number of moles of gas in a given sample does not change with temperature or pressure.

Summary of Gas Laws: The Ideal Gas Equation

Any sample of gas can be described in terms of its pressure, temperature (in kelvins), volume, and the number of moles, n, present. Any three of these variables determine the fourth.

ideal gas: gas that exactly obeys the gas laws

ideal gas equation: $PV = nRT$

R = universal gas constant; its units depend on the choices of units for P, V, and T

$R = 0.0821\ \text{L} \cdot \text{atm/mol} \cdot \text{K}$

Determination of Molecular Weights and Molecular Formulas of Gaseous Substances

The ideal gas equation can be used to determine molecular weights and molecular formulas.

Dalton's Law of Partial Pressures

Dalton's Law of Partial Pressures: the total pressure exerted by a mixture of ideal gases is the sum of the partial pressures of those gases,

$$P_{\text{total}} = P_A + P_B + P_C + \ldots..$$

partial pressure: the pressure that each gas exerts in a mixture of gases mole fraction, X: the number of moles of a component divided by the total number of moles,

$$X_A = \frac{\textit{no. mol A}}{\textit{total number of moles}}$$

$$X_B = \frac{\textit{no. mol B}}{\textit{total number of moles}}$$

and so on,

The sum of all mole fractions in a mixture is equal to 1.

$$P_A = X_A \cdot P_{\text{total}};\ P_B = X_B \cdot P_{\text{total}};$$

and so on,

The partial pressure of each gas is equal to its mole fraction times the total pressure of the mixture. A gas in contact with water soon becomes saturated with water vapour. The pressure inside the container is the sum of the partial pressure of the gas itself plus the partial pressure exerted by water vapour in the gas mixture. Every liquid shows a characteristic vapour pressure that varies only with temperature.

Mass-Volume Relationships in Reactions Involving Gases

One mole of a gas, measured at STP, occupies 22.4 L; we can use the ideal gas equation to find the volume of a mole of gas at any other conditions. This information can be utilised in stoichiometry calculations. Gay-Lussac's Law of Combining Volumes: at constant temperature and pressure, the volumes of reacting gases can be expressed as a ratio of simple whole numbers The ratio is obtained from the coefficients in the balanced equation for the reaction.

The Kinetic-Molecular Theory

The basic assumptions of the kinetic-molecular theory for an ideal gas are:

- Gases consist of individual molecules that are very small and very far apart.
- Gas molecules are in continuous, random, straight-line motion.
- Collisions between gas molecules and with the walls of the container are perfectly elastic.
- Gas molecules exert no attractive or repulsive forces on one another.

Diffusion and Effusion of Gases

- *Effusion:* The escape of a gas through a tiny hole or a thin porous wall
- *Diffusion:* The movement of a substance into a space or the mixing of one substance with another.

Because they move faster, lighter gas molecules effuse through the tiny openings of porous materials faster than heavier molecules.

Deviations from Ideal Gas Behaviour

Nonideal gas behaviour is most significant at high pressures and/or low temperatures.

THE GAS SPECIFIC HEATS CV AND CP

Consider now the two specific heats of this same sample of gas, let's say one mole:

Specific heat at constant volume, C_V (piston glued in place),

Specific heat at constant pressure, C_P (piston free to rise, no friction).

In fact, we already worked out C_V in the Kinetic Theory lecture: at temperature T, recall the average kinetic energy per molecule is $\frac{3}{2}kT$, so one mole of gas—Avogadro's number of molecules—will have total kinetic energy, which we'll label internal energy,

$$E_{\text{int}} \tfrac{3}{2}kT.N_A \tfrac{3}{2}RT.$$

(In this simplest case, we are ignoring the possibility of the molecules having their own internal energy: they might be spinning or vibrating—we'll include that shortly).

That the internal energy is $\frac{3}{2}kT$ per mole immediately gives us the specific heat of a mole of gas in a fixed volume,

$$C_V = \tfrac{3}{2}R$$

that being the heat which must be supplied to raise the temperature by one degree.

However, if the gas, instead of being in a fixed box, is held in a cylinder at constant pressure, experiment confirms that more heat must be supplied to raise the gas temperature by one degree. As Mayer realized, the total heat energy that must be supplied to raise the temperature of the gas one degree at constant pressure is $\frac{3}{2}k$ per molecule plus the energy required to lift the weight.

The work the gas must do to raise the weight is the force the gas exerts on the piston multiplied by the distance the piston moves. If the area of piston is A, then the gas at pressure P exerts force PA. If on heating through one degree the piston rises a distance Δh, the gas does work

$$PA.\Delta h = P\Delta V.$$

Now, for one mole of gas, $PV = RT$, so at constant P

$$P\Delta V = P\Delta T.$$

Therefore, the work done by the gas in raising the weight is just $R\Delta T$, the specific heat at constant pressure, the total heat energy needed to raise the temperature of one mole by one degree,

$$C_P = C_V + R.$$

In fact, this relationship is true whether or not the molecules have rotational or vibrational internal energy. (It's known as Mayer's relationship.) For example, the specific heat of oxygen at constant volume

$$C_V(O_2) = \tfrac{5}{2}R.$$

and this is understood as a contribution of $\frac{3}{2}R$ from kinetic energy, and R from the two rotational modes of a dumbbell molecule (just why there is no contribution form rotation about the third axis can only be understood using quantum mechanics). The specific heat of oxygen at constant pressure

$$C_P(O_2) = \tfrac{3}{2}R.$$

It's worth having a standard symbol for the ratio of the specific heats:

$$\frac{C_P}{C_V} = \gamma.$$

ISOTHERMS AND ADIABATS

An ideal gas in a box has three thermodynamic variables: P, V, T. But if there is a fixed mass of gas, fixing two of these variables fixes the third from $PV = nRT$ (for n moles). In a heat engine, heat can enter the gas, then leave at a different stage. The gas can expand doing work, or contract as work is done on it. To track what's going on as a gas engine transfers heat to work, say, we must follow the varying state of the gas. We do that by tracing a curve in the (P, V) plane. Supplying heat to a gas which consequently expands and does mechanical work is the key to the heat engine.

But just knowing that a gas is expanding and doing work is not enough information to follow its path in the (P, V) plane. The route it follows will depend on whether or not heat is being supplied (or taken away) at the same time. There are, however, two particular ways a gas can expand reversibly—meaning that a tiny change in the external conditions would be sufficient for the gas to retrace its path in the (P, V) plane backwards. It's important to concentrate on reversible paths, because as Carnot proved and we shall discuss later, they correspond to the most efficient engines. The two sets of reversible paths are the isotherms and the adiabats.

Isothermal behaviour: The gas is kept at constant temperature by allowing heat flow back and forth with a very large object (a "heat reservoir") at temperature T. From $PV = nRT$, it is evident that for a fixed mass of gas, held at constant T but subject to (slowly) varying pressure, the variables P, V will trace a hyperbolic path in the (P, V) plane. This path, $PV = nRT_1$, say is called the isotherm at temperature T1. Here are two examples of isotherms:

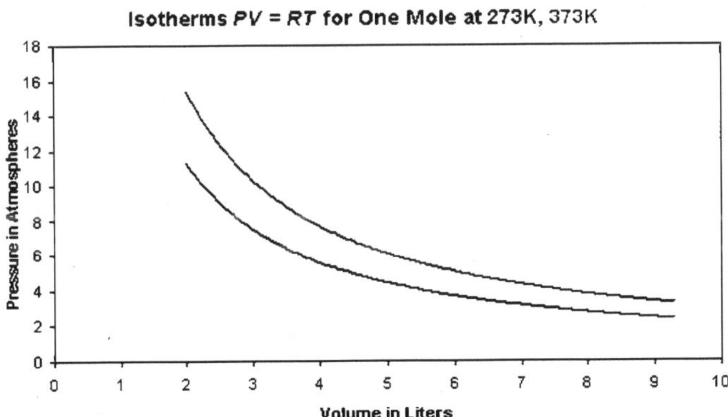

Adiabatic behaviour: "adiabatic" means "nothing gets through", in this case no heat gets in or out of the gas through the walls. So all the work done in compressing the gas has to go into the internal energy E_{int}.

As the gas is compressed, it follows a curve in the (P, V) plane called an adiabat. To see how an adiabat differs from an isotherm, imagine beginning at some point on the blue 273K isotherm on the above graph, and applying pressure so the gas moves to higher pressure and lower volume.

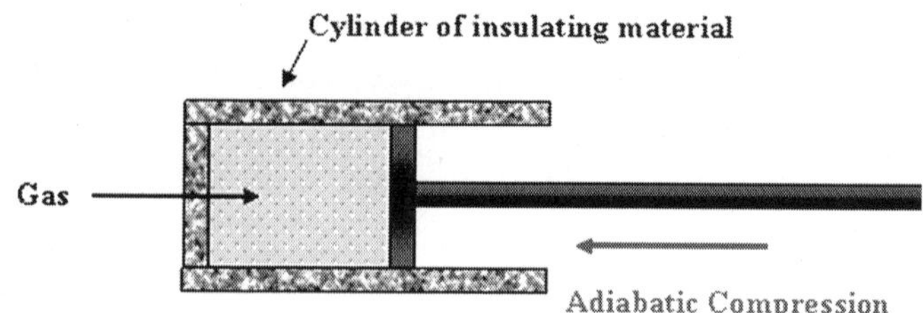

Since the gas's internal energy is increasing, but the number of molecules is staying the same, its temperature is necessarily rising, it will move towards the red curve, then above it.

This means the adiabats are always steeper than the isotherms.

EQUATION FOR AN ADIABAT

What equation for an adiabat corresponds to $PV = nRT_1$ for an isotherm?

On raising the gas temperature by ΔT, the change in the internal energy—the sum of molecular kinetic energy, rotational energy and vibrational energy,

$$\Delta E_{\text{int}} = C_V \Delta T.$$

This is always true: whether or not the gas is changing volume is irrelevant, all that counts in E_{int} is the sum of the energies of the individual molecules (assuming as we do here that attractive or repulsive forces between molecules are negligible). In adiabatic compression, all the work done by the external pressure goes into this internal energy, so

$$-P\Delta V = C_V \Delta T.$$

(Compressing the gas of course gives negative ΔV, positive ΔEint.)

To find the equation of an adiabat, we take the infinitesimal limit

$$-PdV = C_V dT.$$

Divide the left-hand side by PV, the right-hand side by RT (since $PV = RT$, that's OK) to find

$$-\frac{R}{C_V}\frac{dV}{V} = \frac{dT}{T}.$$

Recall now that $C_P = C_V + R$, and $C_P/C_V = \gamma$. It follows that

$$\frac{R}{C_V} = \frac{C_P - C_V}{C_V} = \gamma - 1.$$

Hence

$$-(\gamma - 1)\int \frac{dV}{V} = \int \frac{dT}{T}$$

and integrating

$$\ln T + (\gamma - 1) \ln V = \text{const.}$$

from which the equation of an adiabat is

$$TV^{-1} = \text{const}.$$

From $PV - RT$, the P, V equation for an adiabat can be found by multiplying the left-hand side of this equation by the constant PV/T, giving

$$PV^{\gamma} = \text{const. for an adiabat,}$$

where $\gamma = {}^5/_3$ for a monatomic gas, ${}^7/_5$ for a diatomic gas.

THE CARNOT CYCLE

All standard heat engines (steam, gasoline, diesel) work by supplying heat to a gas, the gas then expands in a cylinder and pushes a piston to do its work. The catch is that the heat and/or the gas must somehow then be dumped out of the cylinder to get ready for the next cycle. We examine the first step, the expansion, then go on to the full cycle—Carnot's analysis. Carnot's aim was to figure out how to maximize the efficiency of a heat engine, and then work out what that efficiency was, that is, how much of the heat supplied was actually converted into the mechanical work done by the engine.

Remember that he had in mind the analogy of the water wheel, at that time still a main driving force of industry. He knew that the most efficient water wheels were those that operated smoothly, the water went into the buckets at the top from the same level, it didn't fall through any height, and didn't splash around. In the limit of a frictionless wheel, with gentle flow on and off the wheel, the machine would be reversible—turning it in reverse to raise the water back would take the same amount of work the wheel had delivered as the water fell. This was clearly perfect efficiency, so these were to conditions to emulate in the heat engine. The analog to having the water flow into buckets at the same height, with no wasteful drop, is to have the heat from the heat supply flow into the gas at the same temperature.

There must of course be a slight drop in temperature for the heat to flow at all, but this must be minimized. This means that as the heat is supplied and the gas expands, the temperature of the gas stays the same as that of the heat supply (the "heat reservoir") and the gas is expanding isothermally.

ISOTHERMAL EXPANSION

So the first question is: how much work is done by an isothermally expanding gas? Taking the temperature of the heat reservoir to be TH (H for hot), the expanding gas follows the isothermal path $PV = nRT_H$ in the (P, V) plane.

The work done by the gas in a small volume expansion ΔV is just $P\Delta V$, the area under the curve.Hence the work done in expanding isothermally from volume Va to Vb is the total area under the curve between those values,

$$\text{work done isothermally} = \int_{V_a}^{V_b} PdV = \int_{V_a}^{V_b} \frac{nRV_H}{V} dV = nRT_H \ln\frac{V_b}{V_a}.$$

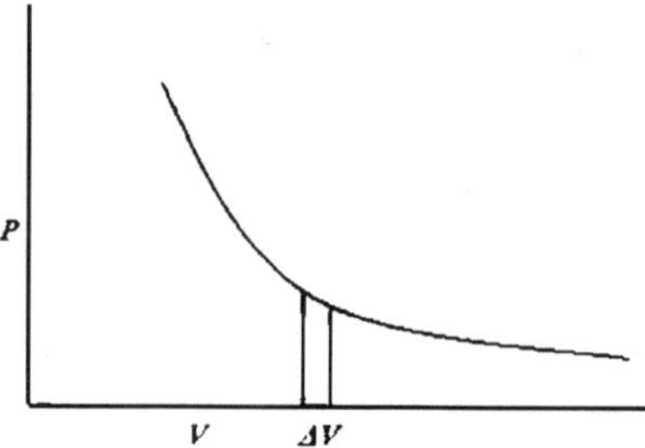

Since the gas is at constant temperature TH, there is no change in its internal energy during this expansion, so the total heat supplied must be $nRT_H \ln \frac{V_b}{V_a}$, the same as the external work the gas has done.

In fact, this isothermal expansion is only the first step: the gas is at the temperature of the heat reservoir, hotter than its other surroundings, and will be able to continue expanding even if the heat supply is cut off.

To ensure that this further expansion is also reversible, the gas must not be losing heat to the surroundings. That is, after the heat supply is cut off, there must be no further heat exchange with the surroundings, the expansion must be adiabatic.

ADIABATIC EXPANSION

The work done in an adiabatic expansion is like that done in allowing a compressed spring to expand against a force—equal to the work needed to compress the spring in the first place, for a perfect spring, and an adiabatically enclosed gas is essentially perfect in this respect. In other words, adiabatic expansion is reversible.

To find the work the gas does in expanding adiabatically from Vb to Vc, say, the above analysis is repeated with the isotherm $PV = nRT_H$ replaced by the adiabat $PV^\gamma = P_b V_b^\gamma$,

work done adiabatically W_{adiabat}

$$\int_{V_a}^{V_c} PdV = P_b V_b^\gamma \int_{V_b}^{V_c} \frac{dV}{V^\gamma} = P_b V_b^\gamma \frac{V_c^{1-\gamma} - V_b^{1-\gamma}}{1-\gamma}.$$

Again, this is the area under the curve, in this case under the adiabat, from b to c in the (P, V) plane.

Since points b, c are on the same adiabat, $P_b V_b^\gamma = P_c V_c^\gamma$, and the expression can be written more neatly:

$$W_{\text{adiabat}} = \frac{P_c V_c - P_b V_b}{1-\gamma}.$$

This is a useful expression for the work done since we are plotting in the (P, V) plane, but note that from the gas law $PV = nRT$, the numerator is just

$nR(Tc - T_b)$, and from this $W_{\text{adiabat}} = nC_V\,(T_c - T_b)$, as of course it must be—this is the loss of internal energy that has been expended by the gas on expanding against external pressure.

We've looked in detail at the work a gas does in expanding as heat is supplied (isothermally) and when there is no heat exchange (adiabatically). These are the two initial steps in a heat engine, but it is equally necessary for the engine to get back to where it began, for the next cycle. The general idea is that the piston drives a wheel, which continues to turn and pushes the gas back to the original volume.

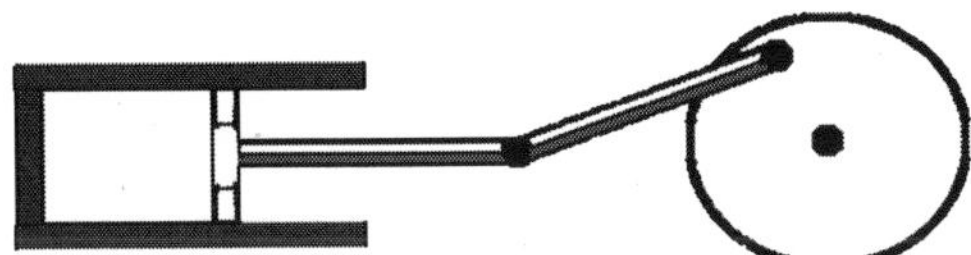

But it is also essential for the gas to be as cold as possible on this return leg, because the wheel is now having to expend work on the gas, and we want that to be as little work as possible—it's costing us. The colder the gas, the less pressure the wheel is pushing against.

To ensure that the engine is as efficient as possible, this return path to the starting point (P_a, V_a)must also be reversible. We can't just retrace the path taken in the first two legs, that would take all the work the engine did along those legs, and leave us with no net output. Now the gas cooled during the adiabatic expansion from b to c, from TH to TC, say, so we can go some distance back along the reversible colder isotherm TC. But this won't get us back to (P_a, V_a), because that's on the TH isotherm.

The simplest option—the one chosen by Carnot—is to proceed back along the cold isotherm to the point where it intersects the adiabat through a, then follow that isotherm back to a. (One could follow a more complicated path: provided it was composed of segments each being adiabatic or isothermal, it would be reversible.) Carnot's cycle is around that curved quadrilateral having these four curves as its sides.

Efficiency of the Carnot Engine

In a complete cycle of Carnot's heat engine, the gas traces the path abcd. The important question is: what fraction of the heat supplied from the hot reservoir (along the red top isotherm) is turned into mechanical work? This fraction is called the efficiency of the engine.

The work output along any curve in the (P, V) plane is just $\int PdV$ —the area under the curve, but it will be negative if the volume is decreasing! So the work done by the engine during the hot isothermal segment is the area abfh, then the adiabatic expansion adds the area bcef, but as the gas is compressed back, the wheel has to do work on the gas equal to the area cdge as heat is dumped into the cold reservoir, then dahg as the gas is recompressed to the starting point. The bottom line is that the total work done by the gas is the area bounded by the four paths: the curved "parallelogram" in the picture above. We could compute this area by finding $\int PdV$ for each segment, but that is unnecessary—on completing the cycle, the gas is back to its initial temperature, so has the same internal energy. Therefore, the work done by the engine must be just the difference between the heat supplied at TH and that dumped at TC.

Now the heat supplies along the initial hot isothermal path ab, equal to the work done along that leg, is (from the paragraph above on isothermal expansion):

$$Q_H = nRT_H \ln\frac{V_b}{V_a}$$

and the heat dumped into the cold reservoir along cd is

$$Q_C = nRT_C \ln\frac{V_c}{V_d}.$$

The difference between these two is the net work output. This can be simplified using the adiabatic equations for the other two sides of the cycle:

$$T_H V_b^{\gamma-1} = T_C V_c^{\gamma-1}$$
$$T_H V_a^{\gamma-1} = T_C V_d^{\gamma-1}.$$

Dividing the first of these equations by the second,

$$\left(\frac{V_b}{V_a}\right) = \left(\frac{V_c}{V_d}\right)$$

and using that in the preceding equation for QC,

$$Q_C = nRT_C \ln\frac{V_a}{V_b} = \frac{T_C}{T_H} Q_H.$$

The work done can now be written simply:

$$W = Q_H - Q_C = \left(1 - \frac{T_C}{T_H}\right) Q_H.$$

Therefore the efficiency of the engine, defined as the fraction of the ingoing heat energy that is converted to available work, is

$$\text{efficiency} = \frac{W}{Q_H} = 1 - \frac{T_C}{T_H}.$$

These temperatures are of course in degrees Kelvin, so for example the efficiency of a Carnot engine having a hot reservoir of boiling water and a cold reservoir ice cold water will be 1– (273/373) = 0.27, just over a quarter of the heat energy is transformed into useful work.

After all the effort to construct an efficient heat engine, making it reversible to eliminate "friction" losses, etc., it is perhaps somewhat disappointing to find this figure of 27% efficiency when operating between 0 and 100 degrees Celsius. Surely we can do better than that?

After all, the heat energy of hot water is the kinetic energy of the moving molecules, can't we find some device to channel all that energy into useful work? Well, we can do better than 27%, by having a colder cold reservoir, or a hotter hot one. But there's a limit: we can never reach 100% efficiency, because we cannot have a cold reservoir at $T_C = 0K$, and even if we did after the first cycle the heat dumped into it would warm it up!

The Second Law of Thermodynamics states that we cannot devise an engine, working in a cycle, that simply extracts heat from a hot reservoir and delivers mechanical work.

This means any engine that takes heat and delivers work also dumps out some of the initial heat to a reservoir at a lower temperature.

It's important to note that the First Law of Thermodynamics, the conservation of total energy including heat, would not be violated by an engine that powered a ship by extracting heat energy from the surrounding water. This Second Law is saying something new. And, this Second Law does not follow from the First by logical deduction—it comes (like the First) from experiment and observation.

SPECIFIC HEAT OF AN IDEAL GAS

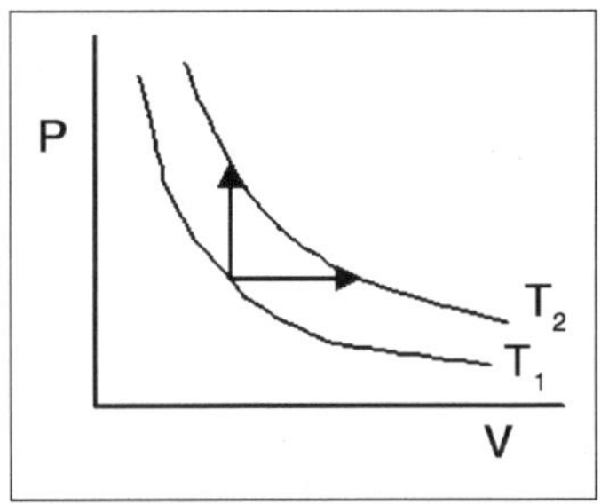

For gases, it is conventional to refer to the *molar* specific heat, which is the heat absorbed per *mole* per unit temperature rise. The molar specific heat, and thus the heat absorbed for a given temperature rise, depends on whether the pressure or the volume is held constant as the temperature increases.

$$Q = nc_v\Delta T \text{ (Constant volume)}$$
$$Q = nc_p\Delta T \text{ (constant pressure)}$$

From the 1st law of thermodynamics, no work is done in a constant volume process. Thus,

$$\Delta U = Q \text{ (constant volume)}$$

For a *monatomic* gas the internal energy is the total translational kinetic energy. Then,

$$U = E = \frac{3}{2}nRT\,, \Delta U = \frac{3}{2}nR\Delta T\,, \text{ and}$$

$$\frac{3}{2}nR\Delta T = nc_v\Delta T \text{, or}$$

$$c_v = \frac{3}{2}nR$$

For a constant pressure process, $\Delta W = -P\Delta V = -nR\Delta T$. So,

$$\Delta U = Q + W = Q - P\Delta V = Q - nR\Delta T,$$

or,

$$Q = \Delta U + nR\Delta T = \frac{3}{2}nR\Delta T + nR\Delta T = \frac{5}{2}nR\Delta T.$$

Then,

$$c_p = \frac{Q}{n\Delta T} = \frac{5}{2}nR$$

and,

$$c_p = c_v + R$$

During a constant volume process, no work is done so all the heat absorbed goes into increasing the internal energy and the temperature. During a constant pressure process, the gas expands and reduces the internal energy by doing work. Thus, more heat can be absorbed for a given temperature change.

SPECIFIC HEAT OF DIATOMIC GASES

Previously, it was shown that,

$$\frac{1}{2}m\overline{v^2} = \frac{3}{2}k_BT$$

Since $\frac{1}{2}m\overline{v^2} = \frac{1}{2}m\overline{v_x^2} + \frac{1}{2}m\overline{v_y^2} + \frac{1}{2}m\overline{v_z^2}$, this means that the average kinetic energy is $\frac{1}{2}k_BT$ per term or per degree of freedom. A diatomic molecule is somewhat like two masses connected by a spring. The molecule can rotate about its centre of mass and the atoms can vibrate back and forth along the line connecting them. Since the rotational inertia about an axis connecting the atoms is extremely small, there are two energy terms corresponding to rotation about the other two perpendicular axes (two degrees of freedom). The average kinetic energy of each of these terms is $\frac{1}{2}k_BT$ and the total average rotational energy is k_BT.

The vibrational energy of the mass-spring system consists of a kinetic energy and a potential energy term (also two degrees of freedom). Each of these terms has an average of $\frac{1}{2}k_BT$ and the total average vibrational energy is k_BT.

Because of these additional degrees of freedom, the specific heat of a diatomic molecule is greater than that of a monatomic molecule (½ R for each degree of freedom). The rotational and vibrational energies of a molecule are quantised. That is, only certain discrete energies are allowed. In order to excite these rotational and vibrational levels, the gas must be at a sufficiently high temperature. At low temperatures, only translational motions can occur. At higher temperatures, the molecular collisions are sufficient to excite the rotational levels. At still higher temperatures, the vibrational energy levels can be excited. Thus, the specific heat of a diatomic molecule will increase with temperature, giving evidence of the quantum nature of the energies.

KINETIC ENERGY AND MASS FOR VERY FAST PARTICLES

Let's first think about the kinetic energy of a particle traveling close to the speed of light. If a force F acts on the particle accelerating it in its direction of motion, and the particle moves through a distance d in the direction the force is pushing, the force does work Fd, and this must go into the kinetic energy of the particle (assuming as always conservation of energy).

As a warm up, recall the elementary derivation of the kinetic energy $\frac{1}{2}mv^2$ of an ordinary *non-relativistic* (i.e. slow moving) object of mass m. This can be done by accelerating the mass with a constant force F, and finding the work done by the force (force distance) to get it to speed v from a standing start.

The kinetic energy of the mass,

$E = \frac{1}{2}mv^2$,

equals the work done by the force in bringing the mass up to that speed from rest: if it takes time t, and the (uniform) acceleration is a, then the final speed $v = at$, the average speed during acceleration is $\frac{1}{2}v$, so the distance traveled is $\frac{1}{2}vt$, the work done by the force is

$\frac{1}{2}Fvt = \frac{1}{2}mavt = \frac{1}{2}mv^2$.

It is easy to extend this to show that if the mass was initially moving at speed u, the work done by the force is

$$\frac{1}{2}mv^2 - \frac{1}{2}mu^2.$$

We are now ready to think about the change in kinetic energy of a particle acted on by an accelerating force when the particle is already moving practically at the speed of light, like particles in modern accelerators. The most straightforward way to see what happens is to use Newton's Second Law, which in the form

Force = rate of change of momentum

is still true, but close to the speed of light the speed changes negligibly as the force continues to work-instead, the mass increases. Therefore to an excellent approximation,

Force = (rate of change of mass) c

where as usual c is the speed of light.

To give a concrete example, let us imagine we have a particle moving very close to the speed of light, and we apply an accelerating force F as the particle travels in a straight line through a distance d. (In a particle accelerator, for example, the particle with charge q could enter a region of uniform electric field E and feel a force $F = qE$.)

Then the increase in kinetic energy ∇E is given by

$$\Delta E = Fd = \frac{\Delta m}{\Delta t} cd$$

where we have approximated the rate of change of mass by the actual change over the distance d, assumed small, divided by the time taken to traverse d,

$$\nabla \text{ t} = \text{d/c}$$

Substituting this value for ∇ t in the formula for the increase in kinetic energy ∇ E, we find

$$\nabla \text{ t} = \nabla \text{ mc}^2$$

KINETIC ENERGY AND MASS FOR SLOW PARTICLES

Recall that to get momentum to be conserved in all inertial frames, we had to assume an increase of mass with speed by the factor $1 / \sqrt{1 - v^2 / c^2}$.

This necessarily implies that even a slow-moving object has a tiny mass increase if it is put in motion.

How does this mass increase relate to the kinetic energy? Consider a mass m, moving at speed v, much less than the speed of light. Its kinetic energy

$$E = ½mv^2,$$

as discussed above. Its mass is

$$m / \sqrt{1 - v^2 / c^2}$$

which we can write as $m + dm$, so dm is the tiny mass increase we know must occur. It's easy to calculate dm. For small v, we can make the approximations

$$\sqrt{1 - v^2 / c^2} \approx 1 + \frac{1}{2}\frac{v^2}{c^2}$$

and

$$\frac{1}{1 - \frac{1}{2}\frac{v^2}{c^2}} \approx 1 + \frac{1}{2}\frac{v^2}{c^2}$$

This means the total mass at speed v is $m(1 + \frac{1}{2}v^2/c^2)$, and writing this as $m + dm$, we see the mass increase dm equals $\frac{1}{2}\, mv^2/c^2$. This means that again, the mass increase dm is related to the kinetic energy E by

$$E = (dm)c^2.$$

KINETIC ENERGY AND MASS FOR PARTICLES OF ARBITRARY SPEED

We have shown in the two sections above that when a force does work to increase the kinetic energy of a particle it also causes the mass of the particle to increase by an amount equal to the increase in energy divided by c^2. In fact this result is exactly true over the whole range of speed from zero to arbitrarily close to the speed of light, as we shall now demonstrate. For a particle of mass m accelerating along a straight line under a constant force F,

work done = force distance

so

$$K.E. = I\int Rds$$

Now

$$F = \frac{d}{dt}mv,$$

so

$$K.E. = \int \frac{d}{dt}(mv)ds = \int \frac{ds}{dt}d(mv) = \int vd(mv)$$

$$d(mv) = vdm + mdv = \left(v + m\frac{dv}{dm}\right)dm$$

We find dv/dm:

$$m = \frac{m_0}{\sqrt{1 - v^2/c^2}} \cdot \frac{m_0}{\sqrt{1 - v^2/c^2}} = \frac{v}{c^2 - v^2}m$$

That is,

$$\frac{dm}{dv} = \frac{v/c^2}{1 - v^2/c^2} \cdot \frac{m_0}{\sqrt{1 - v^2/c^2}} = \frac{v}{c^2 - v^2}m$$

so

$$\frac{dv}{dm} = \frac{c^2 - v^2}{mv}$$

and

$$d(mv) = \left(v + \frac{c^2 - v^2}{mv}\right)dm$$

Therefore

$$K.E. = \int v d(mv) = \int c^2 dm = (m - m_0)c^2$$

$$= \left(\frac{m_0}{\sqrt{1 - v^2 / c^2}} - m_0 \right) c^2$$

So we see that in the general case the work done on the body, by definition its kinetic energy, is just equal to its mass increase multiplied by c^2.

To understand why this isn't noticed in everyday life, try an example, such as a jet airplane weighing 100 tons moving at 2,000mph. 100 tons is 100,000 kilograms, 2,000mph is about 1,000 meters per second. That's a kinetic energy $\frac{1}{2}Mv^2$ of $\frac{1}{2}.10^{11}$joules, but the corresponding mass change of the airplane down by the factor c^2, 9.10^{16}, giving an actual mass increase of about half a milligram, not too easy to detect!

m and m_0

We use m_0 to denote the "rest mass" of an object, and m to denote its relativistic mass,

$$\left(\frac{m_0}{\sqrt{1 - v^2 / c^2}} \right).$$

In this notation, we follow French and Feynman. *Krane and Tipler, in contrast, use m for the rest mass*. Using m as we do gives neater formulas for momentum and energy, but is not without its dangers.

One must remember that m is *not* a constant, but a function of speed. Also, *one must remember* that the relativistic kinetic energy is $(m-m_0)c^2$, and *not* equal to $\frac{1}{2}mv^2$, even with the relativistic mass.

Index

A

Absolute temperature 7, 9, 46, 91, 101, 105, 106, 107, 123, 229
Accessible 241
Adiabatic Expansion 21, 274
Adiabatic expansion 21, 22, 23, 75, 76, 274, 275
Adiabatic heating 170
Adiabatic process 5, 78, 85, 136, 149, 176
Adiabatic system 85
Aeronautics 237
Aggregation 260
Air velocity 159
Alkali 60, 61, 62, 63, 64, 65
Ambient level 61, 62
Angular momentum 43, 94
Anionic 58, 59
Apparently 240
Approximately 262, 263, 265
Approximation 234, 246
Arbitrarily 243
Arbitrary 281
Assumptions 233, 243, 263, 264, 269
Astrophysics 238

B

Binomial 243, 244, 245, 246
Boltzmann 231, 232, 233, 234, 238, 255, 259, 263, 264, 265, 266
Boltzmann Distributions 33
Brownian motion 148

C

Calculation of specific heats 74
Carnot Cycle 20, 29, 273
Carnot engine 8, 24, 125, 129, 130, 131, 167, 168, 190, 276
Change in entropy 109, 131, 134
Characteristic 265, 269
Chemical equilibrium 187
Chemical reactions 185
Chemical Thermodynamics 144
Classical thermodynamics 129
Classius statement 163
Coefficients 269
Coefficients of expansion 67
Collisions 230, 231, 232, 233, 263, 264, 269, 279
Concept of irreversibility 170
Condensed 266
Conservation 108, 126, 132, 150, 166, 170, 279
Conservation of energy 4, 5, 13, 77, 78, 124, 151, 162, 227
Constant Pressure 153
Control volume 85, 114, 128, 129, 189, 190
Conversion 3, 78, 105, 106, 162
Criterion 258
Cycles 174, 178

D

Densities 243, 266, 268
Derivation of the Joule 15
Deviations of Actual 11

Deviations of actual 11
Diatomics 237
Distinguishable 253, 254, 255

E

Elementary 241
Energy Equation 155, 157
Enthalpy 58, 155, 207, 208, 238, 239
Entropy 55, 57, 58, 82, 84, 108, 109, 110, 111, 112, 113, 115, 125, 127, 128, 129, 130
Entropy and Cosmology 139
Entropy and Information Theory 137
Entropy Balance Equation 145
Entropy changes 113, 131, 133, 184
Entropy in Astrophysics 135
Entropy in quantum mechanics 135
Equation for an Adiabat 19, 272
Equation of state 41, 44, 67, 68, 69, 70, 71, 73, 75, 118
Equilibrium 1, 3, 5, 8, 10, 12, 33, 41, 55, 57, 58, 59, 63, 64, 66, 67, 77, 86, 88, 90, 96
Equipartition theorem 51, 52, 88, 90, 91, 92, 94, 97, 229
Equivalence of statements 164
Equivalent 241, 253, 254, 265
Essentially 265
Extensive properties 80

F

Fahrenheit 7, 9, 147, 148
Fluctuations 244

G

Gas Law 208
Gas Specific Heats 17, 269
Gas Turbine 155
Gas turbine 84
Gaseous 263, 264, 266, 268
Gases 239
Genuinely 254

H

Harmonic oscillators 88, 97
Heat Capacities 182
Heat capacities 182, 207
Heat Capacity 96, 104
Heat capacity 13, 33, 45, 70, 71, 90, 91, 92, 93, 95, 96, 97, 98, 99, 102, 103, 107, 136
Heat engine 83, 130, 134, 163, 167, 172, 173, 174, 175, 178, 190
Heat engines 16, 20, 151, 152, 273
Heat Sink 160
Heat source 159, 189
Helmholtz 58
Homogeneously 243
Hutchinson 62, 63
Hypothesis 139, 163

I

Ice Melting Example 138
Ideal Gas 10, 16, 148
Ideal gas 8, 10, 11, 12, 14, 18, 29, 44, 45, 46, 47, 48, 52, 66, 68, 69, 70, 72, 73, 74, 75
Ideal monatomic gases 43
Informal Descriptions 163
Internal Energy 153, 154
Internal energy 81, 110, 112, 155, 172, 173
Interpreted 257, 260
Intrinsic 258
Irreversibility 231
Irreversible Processes 139
Isolated system 82, 125, 127, 132, 173
Isotherm 10, 11, 12, 18, 19, 20, 21, 22, 23, 30,
Isothermal and adiabatic 75
Isothermal process 176

J

Joule-Thomson 12, 13, 14, 15

K

Kinetic 279, 280, 281, 282

L

Lagrangian 258
Law of Conservation 5
Law of hermodynamics 3
Law of Thermodynamics 6, 7, 24, 29, 30, 128, 139, 147, 151, 152, 168, 277
Law of thermodynamics 1, 3, 14, 42, 45, 68, 72, 75, 77, 78, 90, 98, 105, 113, 119, 123
Linear expansion 148

M

Macroscopic 110, 172, 230
Macrosystems 223, 224, 234
Mathematical formulation 77